The Open University

Arts/Mathematics/Science/Technology
An Inter-faculty Second Level Course in the History of Science

Science and Belief: from Copernicus to Darwin

Block 1 Units 1–3

The 'Conflict Thesis' and Cosmology

The Open University Press

COVER ILLUSTRATION: *Copernicus' own copy of his book* De Revolutionibus Orbium Coelestium. *The hole left by his compass in the left-hand page is visible (from Erich Lessing,* Discoveries of Space, *courtesy of Erich Lessing and Search Press Ltd.).*

The Open University Press,
Walton Hall, Milton Keynes.

First published 1974.

Copyright © 1974 The Open University.

Designed by the Media Development Group of the Open University.

Printed in Great Britain by
MARTIN CADBURY, A SPECIALIZED DIVISION OF SANTYPE INTERNATIONAL,
WORCESTER AND LONDON.

ISBN 0 335 00781 3

This text forms part of an Open University course. The complete list of units in the course appears at the end of this text.

For general availability of supporting material referred to in this text, please write to the Director of Marketing, The Open University, P.O. Box 81, Walton Hall, Milton Keynes, MK7 6AT.

Further information on Open University courses may be obtained from the Admissions Office, The Open University, P.O. Box 48, Walton Hall, Milton Keynes, MK7 6AB.

Contents

Set Books

C. C. Gillispie, *Genesis and Geology*, Harper and Row.

D. C. Goodman (ed.), *Science and Religious Belief: A Selection of Primary Sources*, John Wright and Sons/The Open University Press (ANTHOLOGY I).

R. Hooykaas, *Religion and the Rise of Modern Science*, Scottish Academic Press/ Chatto and Windus.

C. A. Russell (ed.), *Science and Religious Belief: A Selection of Recent Historical Studies*, University of London Press/The Open University Press (ANTHOLOGY II).

Unit 1
Some Approaches to the History of Science

Prepared by Colin A. Russell for the Course Team

ILLUSTRATION OVERLEAF: *Title pages from E. J. Dijksterhuis,* The Mechanization of the World Picture, *Oxford University Press, 1961 edition; J. W. Draper,* History of the Conflict between Religion and Science, *1875 edition, courtesy of Gregg International; J. Y. Simpson,* Landmarks in the struggle between Science and Religion, *Hodder and Stoughton Ltd., 1925; contents pages from* Annals of Science, *courtesy of Taylor and Francis Ltd.;* Isis, *courtesy of the History of Science Society Inc.*

Contents

1 Introduction

This course is about the interactions between what we may call 'scientific' and 'non-scientific' belief-systems. More accurately we should stress that it is about a small though important part of these interactions; small because it relates to a mere four centuries of recorded time and chiefly to one area of the inhabited world (Western Europe); important because the rise of science during this period constituted a development of major importance for the human species, while at the same time profoundly influencing man's beliefs about a vast range of things outside science itself.

The sub-title makes it immediately clear that this relationship will be treated historically. Of course, this is not the only, and certainly not the oldest, way of looking at such a subject. For example, it is perfectly possible to study the subject in purely contemporary terms and many people have tried to do just this from the eighteenth century on; the fact that it is almost impossible to say anything worthwhile without some reference to past events has not precluded the attempt from being made. The limitations of such an approach spring from the very nature of the subject itself. Human beliefs, in science and elsewhere, do not stand still. They are constantly changing and do so through mutual stimulus. Hence any interaction between them is essentially dynamic rather than static. It makes no more sense to exclude the historical element from this kind of inquiry than it does from a discussion on contemporary music, art, law, philosophy, theology or any other area of human concern which has changed and developed over the past centuries.

From the limitation of material implied by the course sub-title it must be obvious that the term 'belief' has primarily a religious, and indeed a Christian, connotation. This is not a matter of partisan selection; it is simply a fact of historical reality that, during this crucial and formative period for Western science, it was Christian belief that it chiefly encountered at every stage of its development and with which it reacted in a great variety of significant ways. To be sure, other aspects of Western intellectual thought impinged upon science from time to time, and these will be noted in their proper place. But above all it was organized Christianity, as represented by the Roman Catholic and Protestant churches, that provided, paradoxically enough, both the bitterest opposition *and* the ultimate inspiration for the new science. The relationship continues today and will gain much from the kind of historical examination that this course is trying to offer. We do not wish to imply that the intention is simply to use history as a kind of tool to enable us to penetrate more deeply into the science/religion relationship (though there is nothing wrong with that in itself). Certainly we hope that, at the end, you will have found that historical studies have helped to illuminate a considerable part of the contemporary problem. But there is a sense in which history is best studied for its own sake, and we make no apology for presenting this as a course on the *history of science*. Whether or not you happen to have a prior interest in religion or religious matters you should certainly find this material highly relevant to an understanding of the way in which science has arrived at its present form.

Now it must be obvious that the phrase 'science and religion' conjures up for many people a very definite kind of impression. There is often a subconscious association of ideas which may colour all one's thinking. You may care to test this out for yourself in the following question.

1	Clash	11	Hostility
2	Compatibility	12	Incompatibility
3	Conflict	13	Indebtedness
4	Crisis	14	Inspiration
5	Cross-fertilization	15	Marriage
6	Divorce	16	Partnership
7	Enrichment	17	Progress
8	Fight	18	Reconciliation
9	Growth	19	Separation
10	Harmony	20	Struggle

Go through these as quickly as possible, underlining those that seem the most appropriate.

If your response to this question was fairly immediate and direct, you will probably have selected numbers, 1, 3, 8, 11 and 20 (and perhaps 4). If so, your off-the-cuff reactions are entirely typical and have a very respectable ancestry. It has for long been customary to regard religion and science as being at loggerheads, and to have always been in that kind of relationship. But when we approach the past with this notion at the back of our minds we are, perhaps unconsciously, creating a certain intellectual framework in which to fit the facts. However hard we try we can never be completely objective, but at least we can recognize what our preconceptions are. In this case some would say that we are creating a 'model' and we shall reconstruct the past on the lines suggested by the model. Engineers might prefer to call it a 'template', but that is to over-emphasize the degree to which it is likely to determine our interpretation.

Thus for many people an historical treatment of science and belief is likely to be based on this particular model which we shall call the *warfare model*. At the end of the unit, and still more at the end of the course, you will be invited to say how adequate this is. Merely from the list (see above) it would be possible to establish other possible alternatives; try to do so now. There is, for example, what could be called the *divorce model*, implying that science and religion became quite separate after a time: items 6, 12 and 19 would apply (together with some of the 'warfare' terms!). Or one could set up the opposite model which we may call 'symbiosis'—mutual dependence and enrichment. This would include items 2, 5, 7, 9, 10, 15, 16 and 17. Yet again a model based on total *separation* might be suggested (12, 19), but if no contact existed one could hardly run a course on the subject! This does not exhaust all the possibilities by a long way, but it does help to show how easy it is to interpret the past in a certain way. Just as Marxist historians have often been guided by the concept of the class struggle, so historians of science may follow an analogous course by concentrating on the science–religion struggle. In each case a model is being created. What really matters is how open we are to alternative interpretations of history as new facts emerge, and how critical we are prepared to be of the conceptual models of ourselves and others. The assertion that the most appropriate model is that concerned with struggle, conflict, warfare, etc. we shall refer to as the 'conflict thesis'.

The intention of this unit is broadly twofold. First, for those who have not worked in the discipline before there is a brief introduction to the subject of the history of science. To some extent this overlaps in content, if not in treatment, with part of Unit 1 for the history of science course AST 281, *Science and the Rise of Technology since 1800*. Those of you who have completed that course, or are currently engaged upon it, may find it possible to omit some of the following two sections. The other intention of this unit is to look at various ways of approaching the question of science and belief within the context of the history of science. In other words we are now concerned with the subject of *historiography*, i.e. the writing of and about history.

Objectives

There are certain detailed objectives associated with this unit and these are given below.

At the end of this unit you should be able to:

1 Identify the main 'landmarks' in the growth of the history of science as a discipline.
2 Differentiate clearly between primary, secondary and tertiary sources.
3 Evaluate the strengths and weaknesses of internalist, externalist and biographical approaches in the history of science.
4 Approach initially various types of writing in this field and assess their reliability.
5 Understand what is meant by 'Whiggish' historical writing, assess its value or otherwise, and recognize it in practice.
6 Recognize different types of historical treatment of the 'science and belief' theme.
7 Form independent critical judgements on the passages quoted in Section 4.
8 Give a fair résumé of the patristic attitudes to science in the third to sixth centuries A.D.
9 Form a preliminary judgement on the applicability of the 'warfare model'.

2 Development of the History of Science

To many people is seems slightly odd to talk of the history of science. History is about the past, and scientists are obviously concerned with the present and indeed the future. One of the most conspicuous features of science is its forward-looking character. Equipment and apparatus rapidly become obsolete and are consigned to the junk-heap, and a similar fate attends many theories and hypotheses that once enjoyed widespread, if not universal, favour. The phlogiston theory in chemistry, the caloric theory of physics, the earth-centred systems of astronomy have for a long time been relegated to the small print of historical introductions in scientific texts, and are now more probably omitted altogether. A literature which abounds in 'Current Awareness' periodicals finds little space for nostalgic reminiscence. In a similar way old scientific textbooks have, at least until very recently, fetched very little in second-hand book-shops.

Yet despite an apparent if superficial incongruity between history and science there is a long tradition of historical studies focused upon either science as a whole or on one of its component disciplines. The fact that people may find it slightly odd tells us less about the essential nature of science than about attitudes of twentieth-century man. What is really at stake here, as we shall try to show, is a philosophy of history which, rarely articulated, has become part of the fabric of our thinking.

The first serious study in the history of science came from those whom today we should call the scientists. Not only were such studies regarded as being beneficial to the progress of the science but, in many cases, were considered an essential part of it. Astronomy offers perhaps the best example. To calculate the positions of a star or planet it has always been necessary to know something of its behaviour in past ages. Recourse to earlier observations gave a continuous link with the past, and medieval astronomy was remarkable for its dependence upon, and unity with, the science of the Greeks. The historical consciousness was if anything increased with the newly available manuscripts in the Renaissance, and writers like Copernicus were profoundly indebted to such sources. No astronomer at that time could fail to be aware of the continuous development of his subject.

With chemistry the situation was different. Its mysterious and complex predecessor, alchemy, was a bizarre mixture of mystical speculation, fraudulent audacity and genuine experimental enquiry. One thing most alchemists were agreed upon was the antiquity of their art, and many were anxious to trace it in an unbroken line as far back as possible, preferably to Moses or even to Adam! It thus became endowed with some appearance of authenticity and the alchemists strove earnestly to discover more of the history of their subject in order that their own researches might be based more completely on the uncorrupted wisdom of antiquity. Indeed, one regrets to note that where this history was not accessible it was not infrequently invented, so intense was their pursuit of this objective. Thus their literature from an early date includes historical documents; an early example was a fifteenth-century work on *The Truth and the Antiquity of the Art of Chemistry*, by Robertus Vallensis.

Perhaps this is one reason why of all the sciences chemistry has always been particularly aware of its own history. Certainly by the seventeenth century numerous histories of chemistry were appearing, such as the important work by Olaus Borrichius, on *The Origin and Progress of Chemistry* (1668). Others have followed in a continuous stream to our own day. But the point that seems especially important about these early histories of chemistry is that they recognized the temporal dimension of the science as important. Distinctions of the kind made today between chemistry and the history of chemistry are less obvious before the nineteenth century.

By the eighteenth century, however, science had made sufficient progress for its past respectability to be taken largely for granted, or alternatively dismissed as irrelevant. There was no longer a psychological need for an impeccable ancestry. But this did not mean that those who worked in chemistry, physics or the other sciences were to lose interest in the historical development of their subjects. Rather was the reverse true, but now the motivation had changed. History was still an integral part of a science, and was now seen to possess an instructive value for the whole subject. For one thing it had a verve and spice all of its own which added to the sheer joy of scientific study. Thus in 1767 Priestley wrote his *The History and Present State of Electricity*, claiming the history presented 'a pleasing spectacle to the human mind', while his later *History and Present State of Discoveries relating to Vision, Light and Colours* (1772) maintained that the historical treatment was *essential* to the effective exposition of the science. Slightly earlier than these was Lalande's *Astronomia* (1764) where history and science were integrated together. But perhaps the best example is given by Thomas Thomson in his *System of Chemistry* (1822 ed.):

Figure 1   Joseph Priestley (Crown Copyright: Science Museum).

Figure 2   Thomas Thomson, engraved by C. Cook from an original by J. G. Gilbert (James Muspratt, Chemistry).

> The object of this work is to exhibit as complete a view as possible of the present state of chemistry; and to trace at the same time gradual progress from its first rude dawnings as a science, to the improved state which it has now attained. By thus blending the history with the science, the facts will be more easily remembered, as well as better understood; and we shall at the same time pay that tribute of respect to which the illustrious improvers of it are justly entitled.

Thus at the beginning of the nineteenth century we find scientific works that were built on a firmly historical foundation. But not all were so constructed. There was already in existence an alternative approach, and that was to confine the history to an (often lengthy) introduction and keep the remainder of the book for an exposition of the 'modern' science. Such was the case with Boerhaave's *Elementa Chemiae* (1732) and a few other works before 1800. It became much more common in the nineteenth century and the synoptic, Thomsonian, approach began to lose general favour. As that century progressed one may detect a general tendency to curtail even the historical introductions (though when Thomas Henry proposed to omit it altogether from his textbook of chemistry he felt it necessary to make a special apology). In chemistry, at least, it became customary in our own century to limit historical matters to brief paragraphs in small print. Quite clearly all this marks a gradual disenchantment with history (or at least the historical method of exposition) by those who have a duty to expound contemporary science.[1]

---

[1] In some subjects there are now signs of a return to a genuinely historical treatment of certain limited areas, but these developments are far too recent for any reasonable assessment to be made of their importance.

For several centuries histories of individual sciences have been appearing as monographs, and separate from ordinary textbooks. By the late nineteenth century there were substantial histories of astronomy, physics and chemistry, some of which are mentioned in Table I.

Table 1

Astronomy

J. S. Bailly	*Histoire de l'Astronomie ancienne*	1775
J. S. Bailly	*Histoire de l'Astronomie moderne*	1778–83
J. B. J. Delambre	*Histoire de l'Astronomie moderne*	1821
R. Grant	*History of Physical Astronomy*	1852

Physics

J. C. Fischer	*Geschichte der Physik*	1801–8
A. Heller	*Geschichte der Physik*	1882–4
J. C. Poggendorf	*Geschichte der Physik*	1879

Chemistry

J. F. Gmelin	*Geschichte der Chemie*	1797–9
T. Thomson	*History of Chemistry*	1830
F. Hoefer	*Histoire de la Chimie*	1842
H. Kopp	*Geschichte der Chemie*	1843–7
A. Wurtz	*Histoire des Doctrines Chimiques depuis Lavoisier jusqu'à nos jours*	1869

By around 1900, then, there were plenty of histories of individual sciences but the history of science no longer enjoyed widespread popularity as an aid to the teaching of science. It was very much an optional extra.

Can you suggest any reasons for the decline in importance of the history of science as an aid in scientific instruction?

Most obvious was the enormous increase in the volume of scientific facts that needed to be taught. The new science took up so much more space and time that something had to go, and the most obvious candidate was the 'Historical Introduction'. Those who wanted to write about the subject had to produce separate full-scale histories. But it seems that there was more to it than the problem of sheer bulk. After all, why was the *history* so expendable, in place of some other aspect of the subject? Perhaps we should take into serious account the changing attitudes to history itself. 'Scientific' history, that is one using a systematic methodology and striving for objectivity, was really a nineteenth-century phenomenon, dating from the great German historian Ranke. The old eighteenth-century indifference to general history—at least in England —was replaced by a new awareness of historical progress and development. History was seen as evolutionary, progress was always forwards. The supreme boast was made by Herbert Spencer: 'It is certain that man must become perfect.' If man was getting better, then so were a lot of other things, science included. Paradoxically this meant that the renewed interest in the past was liable to lead to a certain contempt for it. Past science was no longer quite the glorious heritage it had once seemed to be, and was rather conveniently forgotten in the great march forward. This attitude has been of immense importance in the history of science and we return to it later, but it is noted here as a possible and partial explanation for the changing attitude of some scientists to the history of their subjects. Perhaps we may also detect a third reason, again in England, in the rise of the heuristic method of science teaching. Advocated most strenuously by the chemist H. E. Armstrong, it proposed science by discovery, with a minimum of formal teaching; as Dr W. H. Brock has pointed out, this derived much of its inspiration from the now outmoded 'faculty psychology' of the nineteenth century. The mind was divided into 'faculties' each of which could be separately trained; thus the faculty of observation and deduction could be specifically educated by heuristic courses in science. You learn by making discoveries yourself, preferably in the laboratory. This is a very time-consuming method of education, so history is squeezed out. But also there is no need for it since it no longer has an important rôle. The greater the proportion of practical to theoretical science, the less likely is there to be room for history. This was also true when early science was taught chiefly as a tool for use in medicine or engineering.

All the examples given so far have two things in common. In the first place, they include no histories of science as a whole, only histories of specialist sciences (chemistry, physics, astronomy and so on). Secondly (and in explanation of this) all were written by working scientists (and, one may add, for their scientific colleagues or pupils). Works of this kind have continued to appear up to our own day. Early in the nineteenth century, however, a development occurred which marked a new departure for scientific historiography. That was the appearance in 1837 of the *History of the Inductive Sciences* (3 vols.) by Dr William Whewell (1794–1866), Master of Trinity College, Cambridge. There was much that was noteworthy about both the book and its author. Whewell had been trained both in mathematics and in mineralogy as well as in the classics, but it is chiefly his philosophy, derived alike from Francis Bacon (1561–1626), and from Kant and other German writers, for which he is remembered. It is not unfair to say that his *History* springs from his philosophical interests, and in so far as this is correct we have a new kind of 'input' into the historiography of science. His work had another aspect of novelty as well. It marks the beginning of a genuine *historiography of science* rather than of individual sciences. Admittedly the sciences are on the whole treated separately in his book but it is at least an attempt to go beyond the traditional one-science approach. Whewell's writings had considerable influence on the Continent (more so than in Britain), but it took a long time for others to approach the history of science in this synoptic and rather philosophical way. A boost was given to the process by the rise of positivist philosophy in Paris. This approach to science (of which we will say more in later units) has certain affinities with Whewell's ideas, especially in his optimistic evaluation of the future prospects of science. Later in the century the history of science was adorned by several other names representing this philosophical tradition. Amongst them we may mention the incumbents of the first university chairs in the History of Science. In 1892 Pierre Laffitte (1823–1903), a philosopher and friend of the founder of positivism, Auguste Comte, became Professor of General History of the Sciences in Paris. Three years later, at Vienna, the title of Professor of Philosophy, especially of the History and Theory of Inductive Sciences, was conferred on the physicist Ernst Mach (who is commemorated today by the Mach numbers that relate the speed of a fast-moving object, such as an aircraft or a rocket, to the velocity of sound). His *Science and Mechanics* was his most famous, but by no means only, essay in the history of science.

Figure 3 Dr William Whewell, Master of Trinity College, Cambridge. c.1866 (Mansell Collection).

Clearly these two appointments indicate a shift in emphasis. There is something approaching a recognition of the history of science as a subject in its own right and a new element has been injected that could be loosely called 'philosophical' but which has certain sociological overtones as well. It is worth remembering that the distinction between history and philosophy was less marked on the Continent then than in England.

Perhaps the first man clearly to argue for a general history of science was Paul Tannery (1845–1904), proposed, but never appointed, as successor to Laffitte in Paris. For what was to have been his inaugural lecture he wrote:

> It is obvious that to be a good historian of science it is not enough to be a scientist. In the first place one must have the will to give oneself up to history, that is, to have a taste for it; one must cultivate the historical sense within oneself, which is a sense entirely different from the scientific; and finally one must master a number of special skills which are indispensable aids to the historian while they are of absolutely no value to the scientist who is concerned only for the progress of science.

Figure 4 Paul Tannery (Osiris, Courtesy of Uitgeverij 'De Tempel').

Thus Tannery was proposing that on to the distinctive skills of the scientist should be grafted an attitude of mind that derived essentially from history. That same year he pursued his objective further by organizing, at an international conference of general historians, a sectional meeting for historians of science. This might have been a harbinger of an immediate and fruitful collaboration between the historians and their more specialist colleagues. But for various reasons this was not to be.

Perhaps the first significant gesture from the side of general history came with the publication in 1949 of *The Origins of Modern Science, 1300–1800* by Herbert Butterfield,[1] then Professor of Modern History at the University of Cambridge. The book opens with the following three paragraphs:

> Considering the part played by the sciences in the story of our Western civilisation, it is hardly possible to doubt the importance which the history of science will sooner or later acquire both in its own right and as the bridge which has been so long needed between the Arts and the Sciences.
>
> For a long time the History of Science Committee in Cambridge had wished to see the establishment of a course of teaching which should deal, not merely with the background of science, but with the history of scientific activity itself, and this not merely in occasional episodes but as the study of a continuous process of development. Special difficulties seemed to be encountered in the attempt to secure lectures on a period which could in any sense be called modern; and it was in view of these difficulties that for this introductory session the Committee felt themselves under the necessity of resorting to the mere 'general historian'—or perhaps it struck them that it would be good to test the sincerity of that lip-service which the 'general historian' had so often rendered to the history of science. The following lectures, which were delivered for the Committee in the Lent and Easter Terms of 1948, are reproduced in the hope that they may interest the historian in a little science and the scientist in a little history. . . .
>
> Nobody, of course, will imagine that the mere 'general historian' can pretend to broach the question of the very modern developments in any of the natural sciences, but it is fortunate that in respect of students in both the Arts and the Sciences the supremely important field for the ordinary purposes of education is one more manageable in itself and, indeed, perhaps more in need of the intervention of the historian as such. It is the so-called 'scientific revolution', popularly associated with the sixteenth and seventeenth centuries, but reaching back in an unmistakably continuous line to a period much earlier still. Since that revolution overturned the authority in science not only of the middle ages but of the ancient world—since it ended not only in the eclipse of scholastic philosophy but in the destruction of Aristotelian physics—it outshines everything since the rise of Christianity and reduces the Renaissance and Reformation to the rank of mere episodes, mere internal displacements, within the system of medieval Christendom. Since it changed the character of men's habitual mental operations even in the conduct of the non-material sciences, while transforming the whole diagram of the physical universe and the very texture of human life itself, it looms so large as the real origin both of the modern world and of the modern mentality that our customary periodisation of European history has become an anachronism and an encumbrance. There can hardly be a field in which it is of greater moment to us to see at somewhat closer range the precise operations that underlay a particular historical transition, a particular chapter of intellectual development.

In this extract identify (and underline) passages which indicate Butterfield's views on
(a)   a functional role for the history of science;
(b)   the new kind of history of science envisaged at Cambridge;
(c)   the historical importance of the scientific revolution.

These passages are respectively 'the bridge . . . between the Arts and the Sciences', 'the study of a continuous process of development' and 'it outshines everything since the rise of Christianity and reduces the Renaissance and Reformation to the rank of mere episodes'.

---

[1] Herbert Butterfield, *The Origins of Modern Science, 1300–1800*, G. Bell and Sons Ltd., 1949.

The impact of this work has been immense, but by now that recognition of the history of science of which Butterfield spoke in his first paragraph was beginning to be granted. If any one man was responsible for this state of affairs that man must be George Sarton (1885–1956). Sometimes called the founder of the discipline of the history of science, Sarton would have been remembered for his monumental four-volume *Introduction to the History of Science* which took the narrative up to the fourteenth century. But he also made immense contributions to the organization of his subject, by his bibliographical research, by his *A Guide to The History of Science* (1952) and perhaps above all by his journal *Isis* which still continues as one of the world's leading periodical publications in the history of science. Sarton himself stood in the same positivist tradition as Laffitte and Mach. For him the history of science was:

Figure 5 *George Sarton* (*Courtesy of ISIS, Smithsonian Institution, Washington*).

> not simply what the title implies, a history of our increasing knowledge of the world and of ourselves; it is a story not only of the spreading light but also of the contracting darkness. It might be conceived as a history of the endless struggle against errors, innocent or wilful, against superstitions and spiritual crimes. It is also the history of growing tolerance and freedom of thought. The historian of science must give an example of toleration in admitting the equal claims to other minds than his of the history of art or the history of religion; he should even be ready to admit the anti-historical attitude of the tough-minded technicians.[1]

Sarton led the way to the wider recognition of his subject when the first International Congress in the History of Science took place in 1929. A second congress two years later was attended by a large Soviet delegation. A paper by B. Hessen ('The Social and Economic Roots of Newton's *Principia*') attempted analysis of Newton's world in terms of the dialectical materialism then fashionable.[2] Whatever the merits of this analysis (and it has still to be shown that Hessen's paper had more than a marginal impact on later historiography) the occasion proved to be the point of departure for a number of distinguished left-wing scientists in Britain, and gave new energy to the radical movement here that was to stress the relations between science and society. Perhaps this was the first *practical* consequence arising from the study of the history of science. More importantly it highlights, though it does not originate, the application of yet another input into the subject: a sociological or even a political one.

If the history of science could be said to have achieved respectability as an academic discipline in its own right before the last war, it was not until that conflict was over that it acquired a truly professional status. Now, however, nourished by its contact with science itself, philosophy, history and sociology, it enjoys all the benefits of professional recognition even in Britain (which was noticeably slower than the United States in this respect). In this country it has several internationally acclaimed journals, many flourishing research schools, and is taught in half the universities, including our own.

Today it is still true that the traditional 'scientist–historian', that is the man who has approached history from his specialism and who remains firmly based both ideologically and professionally as a practising scientist, has been joined by a new species of scholar, the professional historian of science. The latter is likely to have begun his studies in one or other of the sciences but has moved over for a variety of reasons to concentrate his energies on the study of the history of science. This has meant he has had in all probability a post-graduate training in the discipline and has certainly had to acquire that sympathetic attitude to history of which Tannery spoke. Clearly in writing the history of science this person is at an advantage compared with his colleague. Yet it is surely going too far to assert, as some of the more vocal representatives of this new discipline have done, that the history of science that they produce must always be better. It is in all probability better by certain canons of historical scholarship but it is by no means certain that it is more appropriate to the

---

[1] G. Sarton, *A Guide to the History of Science*, Waltham, Mass., 1952, p. 11.

[2] This has recently been reprinted in N. I. Bukharin (ed.), *Science at the Cross-roads*, Cass, 1971.

needs of practising scientists or that it is more truly based upon what science really is and has been. It is easily possible to lose touch with the scientific situation if one has left it for some years and much arrogant nonsense has been written about the exclusive claim of the professional historians of science (as contrasted with the enormous scholarship of some of the greatest scientist–historians). Some of the more partisan criticism sounds rather like the mouse lamenting that a lion was deficient because it was unable to squeak. Some of these attitudes you may encounter during your reading connected with this course so it is well to be aware of them. Doubtless the subject is still young enough to have the growing-pains which will no doubt pass away with time. However, it is pleasant to record that a more temperate and tolerant attitude is assumed by the most distinguished representatives of current history of science. The following example was taken from an article by Professor A. G. Debus (Chicago) on 'The History of Chemistry and the History of Science'.

> I have tried to find out that this 'Scientific-History' tradition has different roots than the more recent 'History of Science'. The result of this has often been a lack of communication between scientist-historians and historians of science. It is a problem that was hardly improved by George Sarton's doubts and reservations about the value of the older tradition. My own feeling is that closer co-operation is needed to overcome this gap. Traditional courses in the History of Chemistry taught by chemists, and similar courses in the History of other Sciences, will surely benefit from the connection with programmes in the History of Science which are likely to have a broader outlook towards the sciences as a whole as well as their connection with other intellectual and social currents. Similarly the History of Science *needs* the interest of the scientist-historians. Both parties should actively aid each other to promote the field as a whole.[1]

One significant result of the rise of this subject in the last quarter of a century (and this is important to the present course) is that just as the history of science now draws upon other related disciplines, so it is also enriching them with some of its own insights. For example, many general historians since Butterfield are now much more sympathetic to the subject and are therefore willing to incorporate some of its conclusions in their own general accounts. As Professor Price (Yale) has observed in *The Science of Science*:[2]

> The effect of history of science upon history itself has been manifold and will probably augment as historians change from those who won a university certificate of ignorance in the sciences to those who grow up in the world where modern science and technology are changing drastically whole areas of the globe, altering the bases of the balances of power and affecting the well-being and security as well as the intellectual outlook of all men.

Much of what has been said above about the history of science covers the history of mathematics as well. In fact many people regard the latter as a branch of the history of science, and there is much to be said for this view. In the eighteenth century, however, the history of mathematics was sometimes taken to include the history of science, or at least part of it. Thus valuable accounts of the development of astronomy are to be found in A. G. Kaestner's *Geschichte der Mathematik* (1796–1800) and in J. E. Montucla's *Histoire des Mathématiques* (1758). The marriage between mathematics and physical science was so strong until the nineteenth century that only after mathematics had begun to assert its independence was its history recognized as being perhaps partially separable from that of science.

In the field of intellectual history (history of ideas) the development of scientific concepts has now won a place, and histories of the kind that this course is attempting can be written with all the resources made available by recent scholarship in the history of science and related areas of intellectual history. It remains therefore to suggest some of the approaches to the subject that are now possible, as well as some of the pitfalls and the hazards that await us.

---

[1] A. G. Debus, *Ambix*, 1971, **18**, 177.

[2] M. Goldsmith and A. Mackay (eds.), *The Science of Science*, Souvenir Press, 1964, pp. 198–9.

3 Approaches to the History of Science

3.1 Source material

Let us suppose that you intend to proceed to the study of some aspect of the history of science that greatly interests you but about which you know relatively little. Where would you start? Assuming that you had access to a reasonably good public library your first inquiries would probably be made in one of the big encyclopedias but you could hardly be said to have mastered the subject merely from that kind of acquaintance. Those who write articles for encyclopedias are usually experts in their own right but almost invariably draw upon collections of scholarly articles, textbooks, and other writings by historians. This last selection of works would give far more information than an encyclopedia's potted version could possibly hope to do. But again you are still dependent upon the works of other scholars. It therefore becomes necessary, if you are really to add to the corpus of knowledge and do original research, that you should explore the material contemporary with the subject that you are studying. So there are three kinds of source material. First, there are what we term the *tertiary* sources, such as encyclopedias, textbooks and other writings which are based usually on other people's research. Then there are the *secondary* sources which are later evaluations of primary sources and which include some of the standard textbooks in the subject and range from relatively slim scholarly articles to immense multi-volume learned monographs. Finally we have the real quarry, the *primary* sources, which are best defined as sources which come into existence at the time of the events that are being studied. These primary sources must always be differentiated from the secondary ones. In this course there are two Anthologies, one of which is exclusively primary source-material and the other is secondary source-material. The first[1] is a wide-ranging collection of documents originating during the periods that we shall be studying. They are reproduced (with one exception) in facsimile form so at no time will there be any doubt as to their nature. The other Anthology[2] is a collection of varied articles, book reviews, chapters from books and the like, all produced within the last forty years and all well outside the period that is being studied. They represent the distillation of other men's scholarship and historical research. They are invaluable as guides but ultimately they depend for their reliability upon two things: the interpretative powers of their authors which guide their selection and treatment of their sources, and the authenticity of the primary materials which they have used. Time and time again in historical writing one can settle disputed points not by reference to secondary sources but by going right back to the primary materials, which may have been perennially misquoted (and therefore misrepresented) through unconscious errors on the parts of historians in the intervening periods.

The following two questions were set by Professor Marwick in his introduction to the course AST 281 (Unit 1).[3] We suggest you look at these now even if you have already completed that course.

---

[1] D. C. Goodman (ed.), *Science and Religious Belief: A Selection of Primary Sources*, John Wright and Sons/The Open University Press, 1973 (ANTHOLOGY I).

[2] C. A. Russell (ed.), *Science and Religious Belief: A Selection of Recent Historical Studies*, University of London Press/The Open University Press, 1973 (ANTHOLOGY II).

[3] The Open University (1973) AST 281 *Science and the Rise of Technology Since 1800*, Unit 1, *Introduction to the History of Science*, The Open University Press.

1 J. D. Bernal, *Science and Industry in the Nineteenth Century*, first published 1936.   ☐
2 *Philosophical Transactions*, 1810–1825.   ☐
3 E. A. Parnell, *Applied Chemistry*, 1844.   ☐
4 *Journal of the Chemical Society*, 1840–1850.   ☐
5 D. E. Chandler, *Lighting by Gas*, 1936.   ☐
6 Minutes of Evidence Before Committee on the Gas Light and Coke Companies Bill, *Parliamentary Papers*, 1809.   ☐
7 William Whewell, *History of the Inductive Sciences*, 1837.   ☐
8 Letter from Humphry Davy to his mother, January 1797.   ☐
9 George Gissing's novel, *Born in Exile*, 1892.   ☐
10 A. E. Musson, 'James Nasmyth and the Early Growth of Mechanical Engineering', *Economic History Review*, 1957.   ☐

The most difficult one there, I suspect, is 7. Here we have a case of something which, as can happen, is both secondary source and primary source. Which it is, depends upon what questions you are asking of it, on what context it is being used in. For late eighteenth-century science it is essentially a secondary source; but for ideas about science held in the 1830s it is very clearly a primary source. 10, a learned article written long after the period which is our subject for study, is obviously a secondary source. Novels (9) often give rise to difficulties when treated as source material. Really, a novel must be a primary source or nothing. A historical novel written today (for example one of the excellent ones by Thomas Armstrong) is certainly not a primary source for the period in which it is set: it might be some kind of primary source for Thomas Armstrong's own period (i.e. the twentieth century); it can have no real status as a secondary source for the period in which it is set, since it is not the function of a novelist to observe the critical canons of the historians. At best a historical novel may be a pleasant and illuminating way into a historical period, but it should never be regarded as more than this.

Letters, diaries and other private papers of scientists or technologists.

Minutes and papers of scientific societies and institutions. Journals of scientific societies and institutions (e.g. of the British Association for the Advancement of Science).

Parliamentary or Royal Committees or Commissions of Enquiry on scientific or technological matters.

Newspapers and magazines.

Contemporary science textbooks.

Drawings and photographs.

Legal documents.

Patents.

Surviving apparatus or machinery.

Surviving laboratories.

Autobiographies and reminiscences of scientists and technologists.

Eye witness accounts—e.g. of machinery, etc.

Contemporary biographies of scientists and technologists.

Treatises, theses, and other extended scientific works (e.g. Darwin's *Origin of Species*).

Archive film.

If you have got anything like as many entries as I have, you have already grasped the essential nature of primary sources.

A word should perhaps be added here about tertiary source material, that is material that is at least two stages removed from the contemporary sources. We have already mentioned articles in encyclopedias, but they are far from being the only kind. Sometimes it is hard to differentiate clearly between secondary and tertiary sources, especially if one is unaware of the author's expertise and experience as a historian. It should not be automatically assumed that tertiary sources are always less valuable than secondary; a sound, scholarly essay based on the most reliable secondary sources is always better than a badly researched piece of work that purports to be essentially a secondary source.

It must be admitted that the distinction between secondary and tertiary sources is not widely employed by the general historian. He might not regard tertiary sources as proper 'history' and certainly would not rely on them. However, there is a special reason for the present emphasis. In a subject as polemical as science and belief there is a large body of writing that tries to argue from history the case for or against religion. Such writing is rarely of high scholarship and must be distinguished from genuine secondary material arising from contact with primary sources.

Which of the following are likely to be tertiary and which secondary sources in the history of science?
1   A biographical dictionary.
2   A school history textbook published in 1930.
3   A school chemistry textbook published in 1940.
4   Partington's encyclopedic four-volume *History of Chemistry* published in the 1960s[1].
5   A treatise on theology and science, with a strong historical emphasis, published by a religious publishing house.
6   Brecht's play, *The Trial of Galileo*.
7   F. Sherwood Taylor's *Galileo and the Freedom of Thought*, originally published by the Rationalist Press Association.

Where the exact book is not specified in this list there must always be an element of doubt. Thus it is very unlikely that either 2 or 3 would ever be based directly on primary material and therefore themselves be a genuine secondary source. However, there are exceptions to prove the rule. A number of school chemistry textbooks originating between the wars from E. J. Holmyard could, so far as their history of chemistry is concerned, be justly regarded as secondary sources, for the author was a distinguished historian of chemistry and was able to write from direct experience of the primary material (particularly concerning alchemy). Similarly, modern biographical dictionaries are usually potted summaries of secondary sources and therefore rightly regarded as tertiary in themselves. But some of the most recent ones (e.g. the multi-volume *Dictionary of Scientific Biography* edited by Gillispie) are written by scholars with first-hand experience of the subject-matter and rise, at least in places, to the level of secondary material. One would expect 4, with its great length and colossal number of references, certainly to be a tertiary source. Perhaps you would be confirmed in that view by the slightly tendentious use of the word 'encyclopedic' in connection with it. Yet if anything in this list is a genuinely secondary source Partington certainly is.[2] Of the thousands of papers and books cited Partington himself claims to have personally examined all but the merest handful. So we simply cannot judge by appearances. 5 is sufficiently vague to be either secondary or tertiary but works of this kind are likely to be polemical and are more usually derived from existing secondary material. 6 makes no pretentions to serious historical scholarship and, in so far as it is source material, it must clearly be tertiary. And of course the same applies to other dramatic reconstructions of the past either in literature or on the stage. However, it could be a

---

[1] J. R. Partington, *History of Chemistry* (four vols.), Macmillan, 1961–4.

[2] In fact several times, when discussing recent developments, Partington refers to his own work and experience, and at those points the material becomes a primary source.

primary source for twentieth-century history; equally the first edition of our encyclopedia would have primary status for its own period. 7 is a polemical work from a publisher dedicated to certain clear ideological objectives. Yet Sherwood Taylor was a careful scholar in the history of science and his work bears all the signs of being an authentic secondary source with much of his material derived from first-hand acquaintance.

3.2 Internalist or externalist?

We have seen above that there are three kinds of source material. Those who read the history of science need to be acutely aware of the differences between them. For people who write about the history of science, however, there is another quite different threefold distinction. This refers to the way in which these different kinds of source material may be handled, and in fact the whole approach to the subject-matter. These are questions of emphasis more than anything else and the three approaches are by no means mutually exclusive. They are simply labels that are frequently used in the literature and, although they are merely matters of convenience, if they are to be used at all it is as well to have a clear notion as to their meaning. It is particularly necessary since sometimes they are used in an almost pejorative sense.

3.2.1 The internalist history of science

This is an approach to the subject which focuses attention primarily on the changes in scientific theory and practice, frequently within the confines of a single science only and sometimes only a part of that. Such historical writing is usually theoretical, often abstruse and occasionally almost esoteric in character. But it has been for several centuries the standard form of the history of science. Today a number of historians of science, and probably most general historians, tend to look at it askance and to prefer the externalist approach (see below).

What obvious deficiencies is an internalist history of science likely to have?

The most obvious of these is probably the neglect of external factors which might well be supposed to have played a part in the development of scientific theories and practice. Associated with the concentration on internal scientific development there is often a tendency to see changes in those theories, primarily though not entirely in terms of the internal dynamics of science itself. This point will be elaborated in the next section. The other defect is that of course the more profound the analysis in a thorough-going internalist account the less accessible such a work will be. In other words it must partake of the usual limitations of highly specialist studies.

But it may also have their virtues. Not the least of these could be a careful examination of those intellectual factors which are deeply engrained in scientific thought and which can so easily escape an analysis primarily devoted to the external influences on science. Further, by the very nature of their preoccupation, authors of internalist works offer a rich documentation upon which many other scholars of all persuasions may draw. And until the internal intellectual framework of science is clearly perceived talk of external influences is rather pointless. An excellent example of internalist writing is the *History of Chemistry* by J. R. Partington referred to earlier. This is to be found in most large public libraries and is worth, at the very least, a few minutes' inspection to gain some impression of the bewildering richness of the material. Even Partington's sternest critics have admitted that. Other notable 'internalist' writers include Pierre Duhem (1861–1916) and Alexander Koyré (1892–1964).

Despite the contrary trend, recent scholarship has produced a number of excellent monographs that are chiefly internalist in character. Two notable books are *The Copernican Revolution: Planetary Astronomy in the Development of Western Thought* by T. S. Kuhn (1966), and the *Caloric Theory of Gases: from Lavoisier to Regnault* by R. Fox (1969). Though both authors are aware of external influences on science the nature of their chosen subject necessitated a predominantly internalist approach.

Figure 6   Professor J. R. Partington (Crown Copyright: Science Museum).

22

3.2.2 The externalist history of science

The essence of externalist thinking can be put like this. Science has not developed only, or even chiefly, in response to pressures from within, but is and has been moulded by external forces of various kinds. Ignoring internalist considerations, and for the moment regarding 'Science' as a whole, let us consider how such forces may operate. We can begin with a tentative scheme like this:

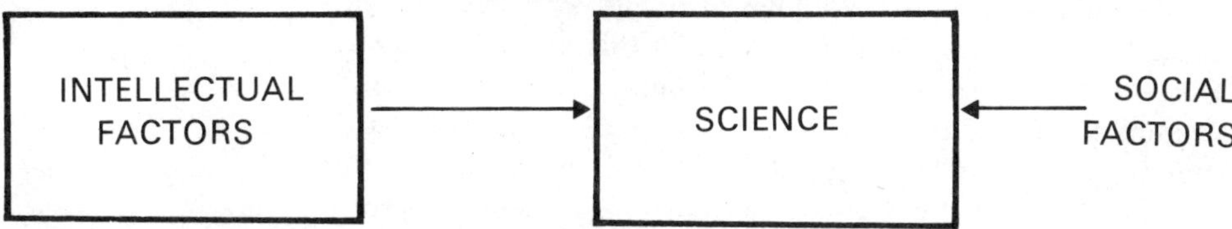

Attempt to elaborate this scheme by identifying individual types of intellectual and social factors.

One would probably emerge with a scheme like this:

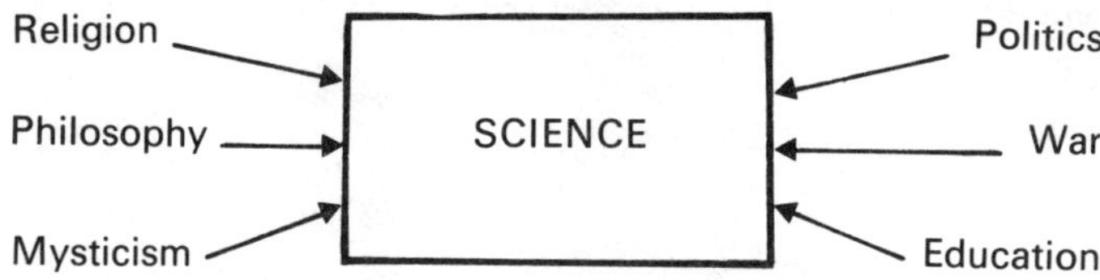

The rather unsatisfactory general term 'mysticism' includes aesthetic and other non-rational elements.

If one now goes on to ask how the social factors could affect science, one might guess that, generally speaking, it would be via technology (which for the moment we can take as 'applied science' though this was certainly not the case earlier than two hundred years ago). So now we have:

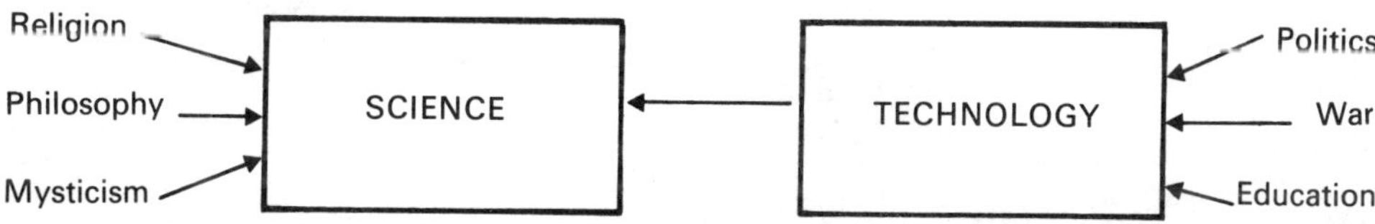

But as everyone knows, science has greatly influenced modern technology so a reverse interaction must be shown as well. And there are cases where political ideologies have directly affected science, as in the famous Lysenko affair, for example. Perhaps political forces are both intellectual and social in character and should be represented on this diagram as a kind of hybrid, as also could education. Thus we find our scheme has to be modified further:

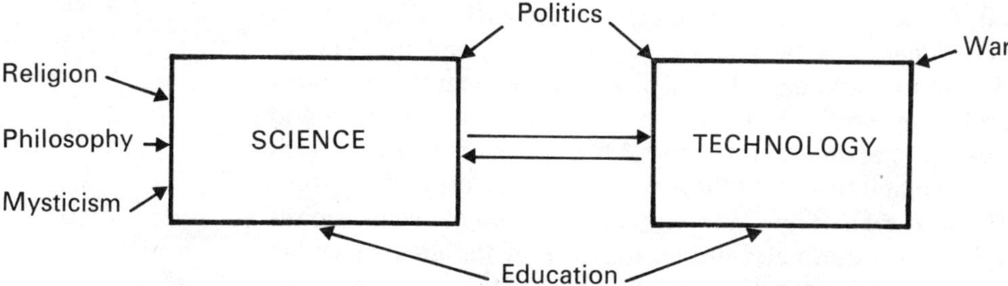

Now simply to write down schemes of this kind does not in the least guarantee that they correspond to anything in reality, either in the present or the past. However, the fact is that recent scholarship has found that such models do make sense and do genuinely appear to be meaningful. The belief that modern science is a product of social pressures does, of course, conform with a Marxist view of history but it would be quite mistaken to identify a general externalist approach with that view, although some historians antagonistic to Marxism have, on that account, been over-cautious in their reluctance to examine external influences upon science and technology.

An important point is that the external factors are by no means limited to those of a social nature but do indeed include intellectual influences. In so far as this is true the history of science partakes of the character of genuine intellectual history, or the history of ideas. Now this was recognized by a few isolated prophets back in the nineteenth century who sought to include in a general history of ideas an account of the development of scientific theories and attitudes. Perhaps the most important of these intellectual histories is that by J. R. Merz, *A History of European Thought in the Nineteenth Century*, of which the first two volumes deal with scientific thought (1904). Historians of science have also seen their discipline in this light. By way of example we may quote the two set books for this course by Gillispie and Hooykaas, together with numerous recent studies of the changing conceptions of matter in the nineteenth century; outstanding amongst these are *Atoms and Elements* by D. M. Knight[1] and *Affinity and Matter* by T. H. Levere[2]. As soon as one gives serious consideration to the external relations of science it becomes clear that a number of these are basically reciprocal relationships. In other words, science not only has its inputs but it also has its outputs.

The following scheme has a number of such 'outputs' indicated by letters. From your own general knowledge you may like to suggest some examples.

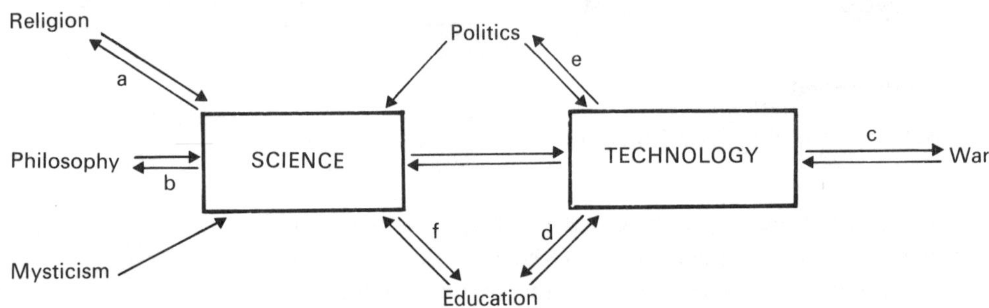

One example for each of these could be:

(a) Impact of Darwinism on Christian belief.

(b) Impact of the scientific revolution on concepts of causality.

(c) Explosives, war-gases and other weaponry.

(d) The Open University!

(e) Public Health.

(f) Educational psychology.

Viewed in this light science is thus seen both as part of a great complex of intellectual issues and at the same time a socially-conditioned phenomenon. The history of science must take into account *both* of these aspects. But even that is not sufficient for, although we have tentatively separated technology from pure science in our discussion, even the word 'science' has a certain ambiguity. Thus it would be wrong to suppose that all that goes under this name is the making, testing and remaking of scientific theories or hypotheses, although this is probably the central and most important part. What about the conditions under which scientists were trained and worked? In what senses was science an organization? Where do scientific institutions come in, if at all? We have to admit that all of these are relevant factors and that they will engage in an interplay with each other. Thus only fairly recently have historians begun to appreciate the importance of roles played by some of the great scientific institutions of the past which have nurtured many a young scientist and in a multiplicity of ways fostered the development of scientific theory. Histories of such notable institutions as the Royal Society, The Lunar Society of Birmingham and many others are beginning to give a new dimension to our understanding of the ways in which science has developed.

[1] D. M. Knight, *Atoms and Elements*, Hutchinson, 1967.

[2] T. H. Levere, *Affinity and Matter*, Oxford University Press, 1971.

3.2.3 A biographical approach

This is a third approach to the history of science which focuses chiefly upon the scientist himself. If you wish to stress the evolution of science as a function of vast, impersonal social forces you are not likely to favour a biographical treatment. Conversely, if you feel that science is a human activity comparable in some respects to music and the arts then you may be predisposed towards this kind of approach. It is important to recognize one's own predispositions and then (if they are merely prejudices) to discard them. Nor is it relevant to good historiography to be over-influenced by the fact that biographies can be fascinatingly exciting or insufferably dull. The crucial question is whether the story of individual lives can tell us anything worth while about the way that science itself has developed. If it can, then it is as worth while to study the lives of Newton and Faraday as it is of Beethoven and Napoleon. Of course, there are dangers in such an approach. It can easily degenerate into fulsome anecdote or it can be merely a featureless list of discoveries that happen to have been made by one man. Perhaps the great men are not always the best to be studied, just because they are great. The minor figures, the third-rate scientists, are often more rewarding because they lacked the genius of the great which separates them from the common man, and because they are likely to reveal more generally accepted scientific attitudes of their day.

The late nineteenth century saw a spate of biographical essays, a good few of them about scientific men. The period witnessed an intensification of what Professor A. R. Hall has called 'The Heroic Sense amongst Scientists'. He writes:

> The French Ministry of Education officially raised Descartes, Pascal, Fermat and Lavoisier to the rank of heroes of the scientific revolution by publishing monumental (and excellent) editions of their literary remains; Favaro did the same single-handed for Galileo; the Dutch created a superb literary sepulchre for Christiaan Huygens; the Poles rediscovered the *Commentariolus* besides the original manuscript of *De Revolutionibus*; and so on. If the heroes did not obviously exist it was necessary to create them by scholarship, as the Russians did with Lomonosov and the Germans (dare one say) with Paracelsus. It is perhaps hardly necessary to add that the notion of industrial progress seemed to enhance further the historical importance of science, and (in its more familiar presentations) strengthened the hagiographical conditions still more. The idea that the minds and the hearts of their heroes, no less than politics or war, and that the private life of a Physicist may be no less edifying than that of a Prince, seems to be one of the more nauseatingly banal pieces of Victoriana.[1]

Having acknowledged this, however, we must add that a biographical sketch, even of a 'great man', does not *have* to be fulsome nor need it be uncritical. Some of the best recent scientific biographies have shown characteristics of both internalist and externalist approaches without losing that basic element of 'sympathy' which is essential for effective biography. Perhaps the most important of these is the biography of Michael Faraday by Pearce Williams. Such works are devoid of hagiography but are not needlessly iconoclastic. They treat the subject as a product of his times and attempt a full portrait, 'warts and all'. Their success highlights a difficulty encountered in other works, the problem of Whiggishness.

---

[1] *Brit. J. Hist. Sci.*, 1969, **4**, 209–10.

3.3 The problem of Whiggishness

It is unlikely that anyone today will read far in the critical literature of the history of science before encountering this problem. Reviewers are fond of applying the word Whiggishness as a term of opprobrium to works which they think display a certain specific deficiency of historical writing. It has nothing to do with the use or misuse of the three types of source material mentioned earlier and, although it is sometimes used with particular reference to 'internalist' writings, it really is of wider application and certainly must not be identified with the productions of the internalist schools.

In 1931 Professor H. Butterfield published his *The Whig Interpretation of History*[1] but only quite recently has the concept been seriously applied in the historiography of science. It is probably true that the more 'professional' the young historian of science feels today the more sensitive he is to this kind of deficiency in the writings of others. As we shall see it is unfashionable to write the kind of history that could be called 'Whiggish', and one must be aware that there are fashions in the history of science as in anything else.

Whiggishness in the history of science is really an attitude of mind which is preoccupied with the triumphant progress of science from its early days to the present, and which looks back upon past achievements through the spectacles of contemporary scientific attitudes. Now in a sense there is nothing wrong with this. Indeed, it may be said that once one has had some acquaintance with modern science it is impossible ever completely to dissociate oneself from it, and so the past is always studied in terms of the present. But it is rather a question of emphasis. The point about this approach which is so much decried by some modern historians of science is that it fails to take into account the *less successful* aspects of scientific progress and it concentrates only on those which led to the present state of affairs. Thus, for example, it largely ignores such strange pseudo-sciences as astrology and phrenology (a Victorian fashion for correlating psychological effects with the shape of one's head!). It never explores the blind alleys or the cul-de-sacs or the strange secondary paths of experience which led ultimately nowhere but which loomed largely in the minds of those who explored them. If one is writing a kind of scientific genealogy, showing how a certain doctrine of science today was derived from a similar one five years ago and that in turn from a yet earlier version, and so on back to an arbitrary starting-point, then one can well afford to ignore such false starts. But at the end of the day one will certainly have no real insight as to why the history turned out as it did. If one wishes to discover something of the underlying reasons for the developments in science, one must certainly take into account the 'bad' as well as the 'good' and the failures as well as the successes.

A Whiggish historiography will of course determine the selection of facts and the whole treatment of them. But it is sometimes easier to detect by certain characteristic thought-forms and turns of phrase. Because of its insistence upon the virtues of present-day science it will tend to regard the science of previous times as less than perfect and certainly as inferior to that contemporary with the writer. As a result certain value-judgements will be made on the workers of an earlier generation and these will be praised or censured according to the effectiveness of their contribution to the mainstream of scientific development. Thus for those who make mistakes there will be reserved censure for the opportunities lost, for their failure to make the correct deductions from the data they had, for inaccuracies in measurement of which they were unaware but with which we happen to be familiar, and so on. On the other hand, those who succeed will be praised, and there may even be animated discussion as to who was exactly the first to arrive at any given conclusion, the point of such debate about priorities being almost only confined to a kind of historical photo-finish: who got there first? In the same way those who arrived at conclusions that bear formal resemblance to scientific beliefs held today will be praised for their 'anticipations' of modern science and for their remarkable prescience.

---

[1] Herbert Butterfield, *The Whig Interpretation of History*, G. Bell and Sons Ltd., 1931.

Now clearly such writing is not really history except in the most limited sense. It is not enough to be able to chart the progress to present perfection; we very much need to know why it took the form it did. It is next to useless to be given merely a chronology of discoveries without any indication as to why they happened when they did, and why some people were able to see the light before others.[1] But the gravest defect of all is that Whiggish writing lacks the element of sympathy that, as we have said before, is essential in historical writing. Perhaps we can do no better than hear the words of a distinguished modern historian of science, Walter Pagel:

> Instead of collecting data that 'make sense' to the acolyte of modern science, the historian should therefore try to make sense of the philosophical, mystical or religious 'side-steps' of otherwise 'sound' scientific workers of the past—'side-steps' that are usually excused by the spirit or rather backwardness of the period. It is these that present a challenge to the historian: to uncover the internal reason and justification for their presence in the mind of the savant and their organic coherence with his scientific ideas. In other words it is for the historian to reverse the method of scientific selection and to restate the thoughts of his hero in their original setting. The two sets of thought—the scientific and the non-scientific—will then emerge not as simply juxtaposed or as having been conceived in spite of each other, but as an organic whole in which they support and confirm each other. There is no other way to lay the savant open to our understanding.[2]

Where is Whiggish history most likely to be found?

It must be obvious from what has been said above that it is a tendency most likely to be found in the writings of the scientist-historians who, naturally, look first at the present state of their subject and only secondly at its antecedents. This is the old tradition, as we have seen, in many branches of science and the 'Whiggish historians' did yeoman service for the subject in those earlier days. But it is very doubtful if today history of science can be regarded as being 'good' if it merely concerns itself with a Whiggish catalogue of discoveries.

The difficulty that faces the scientist-historian whenever he writes is what has been called the 'apocalyptic' character of science itself. One may well see the scientist as a person driven on by the conviction that a 'breakthrough' is inevitable, if not imminent. The present (and near future) is seen as portentous in comparison with the past, and there is a strong temptation to devalue earlier work and so to give an unfair historical assessment. It is certainly true that much recent scholarship has shown how defective are the historical analyses by many distinguished scientists of the past. As Dr John Brooke has observed of a number of recent accounts, 'in almost every case where the history has not been considered superfluous it has been written with a view to establishing the originality of the scientist-historian *qua* scientist'. Note how even Thomas Thomson implies this in the extract on p. 13.

It must not be supposed, however, that scientist-historians are the only ones guilty (if that is the word) of such tendencies. Historians of science who stand in the inductivist tradition of Whewell, the positivists and George Sarton are likely to fall into this way of writing occasionally.

It is a matter that is of particular relevance in this course for, as we shall see shortly, a good deal of Victorian historical writing was of this kind when it came to the matters of 'science and belief'. The nineteenth century was a time when this style of writing was common. Thus in a book entitled *The Story of Chemistry* by H. W. Picton, written at the end of the last century, some of the headings of the periods concerned indicate the extreme Whiggishness of the writer. We have sections on mysticism followed by a period entitled 'The Beginnings of Science', next the 'The Childhood of Truth', then 'The Conflict with Error', leading (of course) to 'The Triumph of Truth'.

[1] This is one of the commonest faults in student essays in this subject. Beware!

[2] W. Pagel, *William Harvey's Biological Ideas*, Hafner, N.Y., 1967, p. 82.

1   Having for many years felt specially interested in Tycho Brahe, it appeared to me that it would be a useful undertaking to apply the considerable biographical materials scattered in many different places to the preparation of a biography which should not only narrate the various incidents in the life of the great astronomer in some detail, but also describe his relations with contemporary men of science, and review his scientific labours in their connection with those of previous astronomers. The historical works of Montucla, Bailly, Delambre, and Wolf have indeed treated of the astronomical researches of Tycho Brahe, but as the plans of these valuable works were different from that adopted by me, I believe the scientific part of the present volume will not be found superfluous, particularly as it is founded on an independent study of Tycho's bulky works.

2   Previous writers have been fascinated by the romantic story of Davy's rather tragic life. My own interest is rather in his scientific work, which I have read many times, seeking to trace the seeds of his greatness and the incidents and accidents that led him astray. Readers will find more chemistry than anecdotes in these pages, and many quotations of Davy's words in an effort to do justice to his scientific work, while trying to explain why, with his great genius, he did not accomplish more.

3   Little research has been done on nineteenth-century scepticism towards atomism, or the frequent outright rejection of Dalton's atomic theory. The present group of studies seeks to examine some of the reasons for this scepticism, and to provide a detailed account of one chemist's serious alternative to atomism.

4   Babbage, Faraday and the rest were growing in an environment in which there were strong restrictive forces preventing the adequate organization of science. If these forces had been removed, the environment would have been different, and so would their achievements. Great though they were, under these conditions they would have been greater still. It might not have been necessary to wait for fifty years for the development of electrical engineering, or a hundred years for the computer. If the British scientists of the last century and a half had been living within an adequate scientific order, Britain today might not have become a second class power. The failure to create an appropriate scientific social order, which could have made full use of the national scientific talent, was due to the inhibiting effect of principles that are inherent in capitalist society.

5   Descartes held the belief that the life process was wholly determined by natural law, that animals were automata without souls and that man alone possessed a rational soul, which influenced his body through one gateway only, that of the pineal gland in the brain (an organ of which the function is still obscure). This view was in sharp contrast to the animism of Stahl, who believed that the soul regulated every part of the life-spirit, and to the views of the Vitalists, who believed that there was Life-force which was capable of performing in the body processes which could not occur outside it. The views of the Vitalists were later weakened by the laboratory synthesis of compounds which were believed to be characteristic of living matter: but Vitalism survives today in a less crude form. Vitalism, in the middle of the nineteenth century, gave way to Materialism, which is the belief that no entity takes part in the life process, other than the atoms, molecules and forces, which constitute and influence non-living matter.

6   In this book we shall survey mathematics primarily to show how its ideas have helped to mould twentieth-century life and thought. The ideas will be in historical order so that our material will range from the beginnings in Babylonia and Egypt to the modern theory of relativity. Some people may question the pertinence of material belonging to earlier historical periods. Modern culture however, is the accumulation and synthesis of contributions made by preceding civilizations. Many of these functions and influences of mathematics are now deeply embedded in our culture.

The first extract comes from the preface to a famous work of the late nineteenth century, a biography of the astronomer Tycho Brahe by J. L. E. Dreyer.[1] There are a few expressions that indicate its Victorian origin (e.g. few people today would use such a phrase as 'review his scientific labours'). But it is very modern in its general approach. Clearly this is a biographical study that is being embarked upon, but the author has a sound historical sense and is trying to relate his subject to contemporary men of science and those of previous ages. It is noteworthy that the author claims to have gone back to the primary sources for his research, though is aware of other secondary sources and has critically evaluated them.

Extract 2 is a quotation from a modern work, another biography, *Humphry Davy*[2] by Sir Harold Hartley. Although this is quite recent (the first edition was 1966) it is in the slightly unfashionable internalist vein as can be realized from the emphasis upon Davy's scientific work and the fact that there is 'more chemistry than anecdotes'. This in fact is a straightforward biography in the time-honoured tradition of history of chemistry by a chemist and for chemists. The approach is clearly biographical and is soundly based on primary sources.

Extract 3 is from another introduction to a modern book, *The Atomic Debates*,[3] edited by W. H. Brock (1967). Unlike the previous one, however, this is well apart from the tradition of rather Whiggish historiography for it deals with one of the notable cul-de-sacs in scientific history. The subject is a somewhat bizarre deviation from atomism and led frankly nowhere. But the study is of great importance for the light it throws on how science really works, how it failed and how it ultimately succeeded in the theory of atomism. This work is in the best tradition of modern historical scholarship.

The fourth extract, also from a fairly modern work, is from the introduction to J. G. Crowther's *Statesmen of Science*[4] (1965). Again a biographical treatment is implied by the title and also suggested by the contents of the paragraphs quoted. But it is clearly biography with a strongly externalist emphasis. Such matters as the organization of science, the use of national scientific talent and the inhibitions which the author attributes to capitalism abundantly confirm that impression. The work derives much from Marx's views of history (though it is well to be reminded again that 'externalist' and 'Marxist' are not synonymous). You may well feel that, in its own way, it is just as Whiggish as some of the writings of the scientist-historian.

The fifth quotation, which comes not from an introduction but from the body of a work, is clearly a highly compressed account of what must be obviously a big subject. Nevertheless the compression is done with skill and confidence, as may be obvious on even the first reading, and the treatment is in the internalist tradition but with a very wide view in which physiology, philosophy, chemistry and biology are all concerned. It is in fact part of Sherwood Taylor's *A Short History of Science*,[5] a fairly recent book which probably comes into the category of a good tertiary source.

The final extract, clearly from a work dealing with the history of mathematics, appears to be (and is) in the old tradition of teaching a scientific or mathematical subject by means of its history.[6] Thus it is not a conventional history of science (or mathematics) for its own sake! That it should appear at all in recent times is perhaps indicative of the different attitudes of mathematicians and natural scientists to the history of their subjects. The short extract quoted reveals an awareness of the cultural effects of mathematics, though it is less specific about the reverse interaction. But one must not jump to conclusions on such partial evidence as this or any other of these short extracts.

---

[1] J. L. E. Dreyer, *Tycho Brahe*, London, 1890, pp. ix–x.

[2] H. Hartley, *Humphry Davy*, S. R. Publishers/E. P. Publishing, 1971, p. vii.

[3] W. H. Brock (ed.), *The Atomic Debates*, Leicester University Press, 1967, p. vii.

[4] J. G. Crowther, *Statesmen of Science*, Cresset Press, 1965, p. 5.

[5] F. Sherwood Taylor, *A Short History of Science*, Heinemann, n.d., p. 203.

[6] Morris Kline, *Mathematics in Western Culture*, Pelican, 1972, p. 16.

4 Historical Treatments of the Science and Belief Theme

It must be obvious from the previous sections that any theme in the history of science can be treated in a multiplicity of ways. This is certainly true of the theme to which this course is devoted. Consequently it is necessary in the remainder of this introductory unit to look at the problems particularly associated with the writing of this kind of history of science. In the sub-sections which follow we are giving fairly lengthy extracts from four authors representing different traditions on historiography and I am asking you to answer questions about each one. To give some kind of coherence to this treatment we are looking at several introductions (or prefaces) in which the authors may be presumed to give some kind of policy statement as to their approach. Extracts that follow from their main texts deal particularly with the views of the early Church Fathers on such problems as the shape of the earth. This subject has the advantage of being fairly compact and a good illustration of the various tendencies in this kind of historical writing. Needless to say the period concerned (the third to sixth centuries A.D.) comes well before the period with which this course attempts to deal. Nevertheless the influence of the Church Fathers (through the so-called Patristic writings) continued for the thousand years that lay between them and the Renaissance in which Copernicus was born. So this section has a twofold aim.

4.1 J. W. Draper's *History of the Conflict between Religion and Science (1875)*

John William Draper (1811–82) was a Lancashire man who became Professor of chemistry and physiology at the University of New York. He was the first President of the American Chemical Society and did notable research on the chemical effects of light. The famous 'Draper effect' in photochemistry was named after him. He laid down the important principle that only light that is absorbed by a substance can give rise to chemical action. Besides writing a treatise on *Human Physiology* he had made several excursions into the historical field, having published also the *History of the Intellectual Development of Europe* and *A History of the American Civil War*. The following quotations come from his famous *History of the Conflict between Religion and Science* published in 1875 and recently (1970) reissued by Gregg International Publishers Ltd. It is a classic in its way and it is highly desirable that you should have at least a nodding acquaintance with it.

The first extract is from the Preface (*op. cit.*, pp. vi–xi).

Figure 7 *John W. Draper (American Chemical Society).*

The history of Science is not a mere record of isolated discoveries; it is a narrative of the conflict of two contending powers, the expansive force of the human intellect on one side, and the compression arising from traditionary faith and human interests on the other.

No one has hitherto treated the subject from this point of view. Yet from this point it presents itself to us as a living issue—in fact, as the most important of all living issues.

A few years ago, it was the politic and therefore the proper course to abstain from all allusion to this controversy, and to keep it as far as possible in the background. The tranquillity of society depends so much on the stability of its religious convictions, that no one can be justified in wantonly disturbing them. But faith is in its nature unchangeable, stationary; Science is in its nature progressive; and eventually a divergence between them, impossible to conceal, must take place. It then becomes the duty of those whose lives have made them familiar with both modes of thought, to present modestly, but firmly, their views; to compare the antagonistic pretensions calmly, impartially, philosophically. History shows that, if this be not done, social misfortunes, disastrous and enduring, will ensue. When the old mythological religion of Europe broke down under the weight of its own inconsistencies, neither the Roman emperors nor the philosophers of those times did any thing adequate for the guidance of public opinion. They left religious affairs to take their chance, and accordingly those affairs fell into the hands of ignorant and infuriated ecclesiastics, parasites, eunuchs, and slaves.

The intellectual night which settled on Europe, in consequence of that great neglect of duty, is passing away; we live in the daybreak of better things. Society is anxiously expecting light, to see in what direction it is drifting. It plainly discerns that the track along which the voyage of civilization has thus far been made, has been left; and that a new departure, on an unknown sea, has been taken. . . .

In thus treating the subject it has not been necessary to pay much regard to more moderate or intermediate opinions, for, though they may be intrinsically of great value, in conflicts of this kind it is not with the moderates but with the extremists that the impartial reader is mainly concerned. Their movements determine the issue.

For this reason I have had little to say respecting the two great Christian confessions, the Protestant and Greek Churches. As to the latter, it has never, since the restoration of science, arrayed itself in opposition to the advancement of knowledge. On the contrary, it has always met it with welcome. It has observed a reverential attitude to truth, from whatever quarter it might come. Recognizing the apparent discrepancies between its interpretations of revealed truth and the discoveries of science, it has always expected that satisfactory explanations and reconciliations would ensue, and in this it has not been disappointed. It would have been well for modern civilization if the Roman Church had done the same.

In speaking of Christianity, reference is generally made to the Roman Church, partly because its adherents compose the majority of Christendom, partly because its demands are the most pretentious, and partly because it has commonly sought to enforce those demands by the civil power. None of the Protestant Churches has ever occupied a position so imperious—none has ever had such wide-spread political influence. For the most part they have been averse to constraint, and except in very few instances their opposition has not passed beyond the exciting of theological odium.

As to Science, she has never sought to ally herself to civil power. She has never attempted to throw odium or inflict social ruin on any human being. She has never subjected any one to mental torment, physical torture, least of all to death, for the purpose of upholding or promoting her ideas. She presents herself unstained by cruelties and crimes. But in the Vatican—we have only to recall the Inquisition—the hands that are now raised in appeals to the Most Merciful are crimsoned. They have been steeped in blood!

What kind of historical writing could be expected from Draper in the light of his own remarks?

Exactly the same points could be made about Draper's way of writing history as were applicable to some of the internal histories of chemistry referred to above. It is written by a 'scientist-historian' and is overwhelmingly Whiggish in approach. Note his use of words and phrases like 'conflict', 'progression', 'daybreak of better things', 'light', etc. and above all the fifth paragraph with its rejection of moderate views. Built into the very title is an enormous assumption, so large that it could easily pass unnoticed: that there was a *conflict* between religion and science and that this was one of the central aspects of the history of science.

How far this 'warfare model' is justified you will be in a position to judge at the end of this course. Further, Draper is clearly writing a polemical work directed against religion in general and Roman Catholicism in particular. Needless to say, his anti-catholic bias has not endeared him to members of the Roman Church; Sir Bertram Windle took particular exception to his book, not least for its caricature of papal infallibility. He wrote in *The Church and Science* (Catholic Truth Society, third edition, 1924):

Perhaps an apology is necessary for raking in that ancient dust-heap of inaccuracies and falsehoods which is entitled 'The Conflict of Religion and Science' and was written by a Dr. Draper, long dead . . . but as this aged work with all its errors intact has recently been published once more with none but an exceedingly obscure indication that it is not the most recent thing in anti-catholicism, it may stand for an example of a grosser form of ignorance on this topic.

However, a standpoint such as Draper's does not automatically invalidate his history; it simply means that we have to take his bias into account when evaluating his historical judgements.

Figure 8   St Augustine
in his cell, by Botticelli
(Uffizi Photo: Mansell
Collection).

The following are now some extracts from Draper's second chapter. Having disposed
of Augustine ('no-one did more than this Father to bring science and religion into
antagonism') he concludes with these remarks on patristic science (*op. cit.*, pp. 63–7).

As to the earth, it affirmed that it is a flat surface, over which the sky is spread like a
dome, or, as St. Augustine tells us, is stretched like a skin. In this the sun and moon and
stars move, so that they may give light by day and by night to man. The earth was made
of matter created by God out of nothing, and, with all the tribes of animals and plants
inhabiting it, was finished in six days. Above the sky or firmament is heaven; in the dark
and fiery space beneath the earth is hell. The earth is the central and most important body
of the universe, all other things being intended for and subservient to it.

As to man, he was made out of the dust of the earth. At first he was alone, but subsequently woman was formed from one of his ribs. He is the greatest and choicest of the
works of God. He was placed in a paradise near the banks of the Euphrates, and was very
wise and very pure; but, having tasted of the forbidden fruit, and thereby broken the
commandment given to him, he was condemned to labor and to death.

The descendants of the first man, undeterred by his punishment, pursued such a career
of wickedness that it became necessary to destroy them. A deluge, therefore, flooded the
face of the earth, and rose over the tops of the mountains. Having accomplished its
purpose, the water was dried up by a wind.

From this catastrophe Noah and his three sons, with their wives, were saved in an ark.
Of these sons, Shem remained in Asia and repeopled it. Ham peopled Africa; Japhet,
Europe. As the Fathers were not acquainted with the existence of America, they did not
provide an ancestor for its people.

Let us listen to what some of these authorities say in support of their assertions. Thus
Lactantius, referring to the heretical doctrine of the globular form of the earth, remarks:
'Is it possible that men can be so absurd as to believe that the crops and the trees on the
other side of the earth hang downward, and that men have their feet higher than their
heads? If you ask them how they defend these monstrosities, how things do not fall
away from the earth on that side, they reply that the nature of things is such that heavy
bodies tend toward the centre, like the spokes of a wheel, while light bodies, as clouds,

smoke, fire, tend from the centre to the heavens on all sides. Now, I am really at a loss what to say of those who, when they have once gone wrong, steadily persevere in their folly, and defend one absurd opinion by another.' On the question of the antipodes, St. Augustine asserts that 'it is impossible there should be inhabitants on the opposite side of the earth, since no such race is recorded by Scripture among the descendants of Adam.' Perhaps, however, the most unanswerable argument against the sphericity of the earth was this, that 'in the day of judgment, men on the other side of a globe could not see the Lord descending through the air.'

It is unnecessary for me to say any thing respecting the introduction of death into the world, the continual interventions of spiritual agencies in the course of events, the offices of angels and devils, the expected conflagration of the earth, the tower of Babel, the confusion of tongues, the dispersion of mankind, the interpretation of natural phenomena, as eclipses, the rainbow, etc. Above all, I abstain from commenting on the Patristic conceptions of the Almighty; they are too anthropomorphic, and wanting in sublimity.

Perhaps, however, I may quote from Cosmas Indicopleustes the views that were entertained in the sixth century. He wrote a work entitled 'Christian Topography,' the chief intent of which was to confute the heretical opinion of the globular form of the earth, and the pagan assertion that there is a temperate zone on the southern side of the torrid. He affirms that, according to the true orthodox system of geography, the earth is a quadrangular plane, extending four hundred days' journey east and west, and exactly half as much north and south; that it is inclosed by mountains, on which the sky rests; that one on the north side, huger than the others, by intercepting the rays of the sun, produces night; and that the plane of the earth is not set exactly horizontally, but with a little inclination from the north: hence the Euphrates, Tigris, and other rivers, running southward, are rapid; but the Nile, having to run up-hill, has necessarily a very slow current.

The Venerable Bede, writing in the seventh century, tells us that 'the creation was accomplished in six days, and that the earth is its centre and its primary object. The heaven is of a fiery and subtile nature, round, and equidistant in every part, as a canopy from the centre of the earth. It turns round every day with ineffable rapidity, only moderated by the resistance of the seven planets, three above the sun—Saturn, Jupiter, Mars—then the sun; three below—Venus, Mercury, the moon. The stars go round in their fixed courses, the northern perform the shortest circle. The highest heaven has its proper limit; it contains the angelic virtues who descend upon earth, assume ethereal bodies, perform human functions, and return. The heaven is tempered with glacial waters, lest it should be set on fire. The inferior heaven is called the firmament, because it separates the superincumbent waters from the waters below. The firmamental waters are lower than the spiritual heaven, higher than all corporeal beings, reserved, some say, for a second deluge; others, more truly, to temper the fire of the fixed stars.'

Was it for this preposterous scheme—this product of ignorance and audacity—that the works of the Greek philosophers were to be given up? It was none too soon that the great critics who appeared at the Reformation, by comparing the works of these writers with one another, brought them to their proper level, and taught us to look upon them all with contempt.

Of this presumptuous system, the strangest part was its logic, the nature of its proofs. It relied upon miracle-evidence. A fact was supposed to be demonstrated by an astounding illustration of something else! An Arabian writer, referring to this, says: 'If a conjurer should say to me, "Three are more than ten, and in proof of it I will change this stick into a serpent," I might be surprised at his legerdemain, but I certainly should not admit his assertion.' Yet, for more than a thousand years, such was the accepted logic, and all over Europe propositions equally absurd were accepted on equally ridiculous proof.

Since the party that had become dominant in the empire could not furnish works capable of intellectual competition with those of the great pagan authors, and since it was impossible for it to accept a position of inferiority, there arose a political necessity for the discouragement, and even persecution, of profane learning. The persecution of the Platonists under Valentinian was due to that necessity. They were accused of magic, and many of them were put to death. The profession of philosophy had become dangerous —it was a state crime. In its stead there arose a passion for the marvelous, a spirit of superstition. Egypt exchanged the great men, who had made her Museum immortal, for bands of solitary monks and sequestered virgins, with which she was overrun.

What is there in this passage which reflects the spirit of the Preface?

Setting aside for a moment the important questions about content and accuracy, we can hardly avoid the impression of an overwhelming, and over-weening, Whiggishness. It is, indeed, a superb example of how *not* to write history. Not only is there an utter lack of sympathy for the people and attitudes being discussed, there is even a cavalier implication that we should 'look upon them all with contempt', while their belief-systems are dismissed as 'presumptuous' and 'this preposterous scheme—the product of ignorance and audacity.'

All of this diatribe tells us a lot about Draper and very little about the history of the period he purports to discuss. There is hardly the slightest attempt to ask *why* such 'preposterous' schemes were launched. To castigate Augustine for arguing that if all men were to see the Second Coming of Christ simultaneously the earth could not be spherical is about as pointless as arguing that he failed to predict also the invention of television which might have made that theoretically possible!

Further criticism of Draper's approach will be found in the section after next. Meanwhile we turn to a near contemporary.

4.2 A. D. White's *A History of the Warfare of Science with Theology in Christendom (1895)*

Andrew Dixon White (1832–1918) was the first President of Cornell University, America's first non-sectarian and co-educational university. He also taught history there, and later was United States Ambassador to Germany and then to Russia. During his early years at Cornell he encountered stiff opposition from various ecclesiastical authorities on account of the liberal policies he was trying to implement. His book arose from that complaint. The following is an extract from his introduction, written while he was ambassador at St Petersburg (*op. cit.*, pp. v–xii).

Figure 9   Andrew Dixon White, c.1880 (Cornell University Archives).

> My book is ready for the printer, and as I begin this preface my eye lights upon the crowd of Russian peasants at work on the Neva under my windows. With pick and shovel they are letting the rays of the April sun into the great ice barrier which binds together the modern quays and the old granite fortress where lie the bones of the Romanoff Czars.
>
> This barrier is already weakened; it is widely decayed, in many places thin, and everywhere treacherous; but it is, as a whole, so broad, so crystallized about old boulders, so imbedded in shallows, so wedged into crannies on either shore, that it is a great danger. The waters from thousands of swollen streamlets above are pressing behind it; wreckage and refuse are piling up against it; every one knows that it must yield. But there is danger that it may resist the pressure too long and break suddenly, wrenching even the granite quays from their foundations, bringing desolation to a vast population, and leaving, after the subsidence of the flood, a widespread residue of slime, a fertile breeding-bed for the germs of disease.
>
> But the patient *mujiks* are doing the right thing. The barrier, exposed more and more to the warmth of spring by the scores of channels they are making, will break away gradually, and the river will flow on beneficent and beautiful.
>
> My work in this book is like that of the Russian *mujik* on the Neva. I simply try to aid in letting the light of historical truth into that decaying mass of outworn thought which attaches the modern world to mediæval conceptions of Christianity, and which still lingers among us—a most serious barrier to religion and morals, and a menace to the whole normal evolution of society.
>
> For behind this barrier also the flood is rapidly rising—the flood of increased knowledge and new thought; and this barrier also, though honeycombed and in many places thin, creates a danger—danger of a sudden breaking away, distressing and calamitous, sweeping before it not only out-worn creeds and noxious dogmas, but cherished principles and ideals, and even wrenching out most precious religious and moral foundations of the whole social and political fabric.
>
> My hope is to aid—even if it be but a little—in the gradual and healthful dissolving away of this mass of unreason, that the stream of 'religion pure and undefiled' may flow on broad and clear, a blessing to humanity. . . .
>
> There was borne in upon me a sense of the real difficulty—the antagonism between the theological and scientific view of the universe and of education in relation to it; therefore

it was that, having been invited to deliver a lecture in the great hall of the Cooper Institute at New York, I took as my subject *The Battlefields of Science*, maintaining this thesis which follows:

In all modern history, interference with science in the supposed interest of religion, no matter how conscientious such interference may have been, has resulted in the direst evils both to religion and to science, and invariably; and, on the other hand, all untrammelled scientific investigation, no matter how dangerous to religion some of its stages may have seemed for the time to be, has invariably resulted in the highest good both of religion and of science. . . .

Meanwhile Prof. John W. Draper published his book on *The Conflict between Science and Religion*, a work of great ability, which as I then thought, ended the matter, so far as my giving it further attention was concerned.

But two things led me to keep on developing my own work in this field: First, I had become deeply interested in it, and could not refrain from directing my observation and study to it; secondly, much as I admired Draper's treatment of the questions involved, his point of view and mode of looking at history were different from mine.

He regarded the struggle as one between Science and Religion. I believed then, and am convinced now, that it was a struggle between Science and Dogmatic Theology. . . .

I close this preface some days after its first lines were written. The sun of spring has done its work on the Neva; the great river flows tranquilly on, a blessing and a joy; the *mujiks* are forgotten.

In what respects do White's objectives appear to resemble Draper's? And how do they seem to differ?

Clearly both Draper and White are alike in their Whiggish philosophy. One may compare the titles of their books. White's whole image of the ice barrier melting in the sunshine of the advancing spring implies an inexorability of progress as well as an uncomprehending attitude to a less favoured past. While understandable in a scientist this comes oddly from a Professor of History. As Arthur Marwick has observed, 'written exclusively from a nineteenth-century point of view, this book showed little of that understanding of a period from the inside which is the characteristic of all good historical writing.'

Of the two writers, Draper was the more hostile to organized religion, while White claims merely to be concerned with the destructive effects of dogmatic theology. Undoubtedly this does mean some difference in emphasis, as White suggests, but in practice it was less than it appears. The citation of Draper (like the rest of the book) suggests that the aspect of religion which intrigued, and horrified, him most was precisely where it erupted into dogmatic theology.

More important differences emerge from a comparison of the main text. The following section from White relates to early views on the form of the earth (*op. cit.*, pp. 91–8).

But, as civilization was developed, there were evolved, especially among the Greeks, ideas of the earth's sphericity. The Pythagoreans, Plato, and Aristotle especially cherished them. These ideas were vague, they were mixed with absurdities, but they were germ ideas, and even amid the luxuriant growth of theology in the early Christian Church these germs began struggling into life in the minds of a few thinking men, and these men renewed the suggestion that the earth is a globe.[1]

[1] The agency of the Pythagoreans in first spreading the doctrine of the earth's sphericity is generally acknowledged, but the first clear and full utterance of it to the world was by Aristotle. Very fruitful, too, was the statement of the new theory given by Plato in the *Timæus*; see Jowett's translation, 62, c. Also the *Phædo*, pp. 449 *et seq.* See also Grote on Plato's doctrine of the sphericity of the earth; also Sir G. C. Lewis's *Astronomy of the Ancients*, London, 1862, chap. iii, section i, and note. Cicero's mention of the antipodes, and his reference to the passage in the Timæus, are even more remarkable than the latter, in that they much more clearly foreshadow the modern doctrine. See his *Academic Questions*, ii; also *Tusc. Quest.*, i and v. 24. For a very full summary of the views of the ancients on the sphericity of the earth, see Kretschmer, *Die physische Erdkunde im christlichen Mittelalter*, Wien, 1889, pp. 35 *et seq.*; also, Eicken, *Geschichte der mittelalterlichen Weltanschauung*, Stuttgart, 1887, Dritter Theil, chap. vi. For citations and summaries, see Whewell, *Hist. Induct. Sciences*, vol. i, p. 189, and St. Martin, *Hist. de la Géog.*, Paris, 1873, p. 96; also, Leopardi, *Saggio sopra gli errori popolari degli antichi*, Firenze, 1851, chap. xii, pp. 184 *et seq.*

A few of the larger-minded fathers of the Church, influenced possibly by Pythagorean traditions, but certainly by Aristotle and Plato, were willing to accept this view, but the majority of them took fright at once. To them it seemed fraught with dangers to Scripture, by which, of course, they meant *their interpretation* of Scripture. Among the first who took up arms against it was Eusebius. In view of the New Testament texts indicating the immediately approaching end of the world, he endeavoured to turn off this idea by bringing scientific studies into contempt. Speaking of investigators, he said, 'It is not through ignorance of the things admired by them, but through contempt of their useless labour, that we think little of these matters, turning our souls to better things.' Basil of Cæsarea declared it 'a matter of no interest to us whether the earth is a sphere or a cylinder or a disk, or concave in the middle like a fan.' Lactantius referred to ideas of those studying astronomy as 'bad and senseless,' and opposed the doctrine of the earth's sphericity both from Scripture and reason. St. John Chrysostom also exerted his influence against this scientific belief; and Ephraem Syrus, the greatest man of the old Syrian Church, widely known as the 'lute of the Holy Ghost,' opposed it no less earnestly.

But the strictly biblical men of science, such eminent fathers and bishops as Theophilus of Antioch in the second century, and Clement of Alexandria in the third, with others in centuries following, were not content with merely opposing what they stigmatized as an old heathen theory; they drew from their Bibles a new Christian theory, to which one Church authority added one idea and another another, until it was fully developed. Taking the survival of various early traditions, given in the seventh verse of the first chapter of Genesis, they insisted on the clear declarations of Scripture that the earth was, at creation, arched over with a solid vault, 'a firmament,' and to this they added the passages from Isaiah and the Psalms, in which it declared that the heavens are stretched out 'like a curtain,' and again 'like a tent to dwell in.' The universe, then, is like a house: the earth is its ground floor, the firmament its ceiling, under which the Almighty hangs out the sun to rule the day and the moon and stars to rule the night. This ceiling is also the floor of the apartment above, and in this is a cistern, shaped, as one of the authorities says, 'like a bathing-tank,' and containing 'the waters which are above the firmament.' These waters are let down upon the earth by the Almighty and his angels through the 'windows of heaven.' As to the movement of the sun, there was a citation of various passages in Genesis, mixed with metaphysics in various proportions, and this was thought to give ample proofs from the Bible that the earth could not be a sphere.[1]

In the sixth century this development culminated in what was nothing less than a complete and detailed system of the universe, claiming to be based upon Scripture, its author being the Egyptian monk Cosmas Indicopleustes. Egypt was a great treasure-house of theologic thought to various religions of antiquity, and Cosmas appears to have urged upon the early Church this Egyptian idea of the construction of the world, just as another Egyptian ecclesiastic, Athanasius, urged upon the Church the Egyptian idea of a triune deity ruling the world. According to Cosmas, the earth is a parallelogram, flat, and surrounded by four seas. It is four hundred days' journey long and two hundred broad. At the outer edges of these four seas arise massive walls closing in the whole structure and supporting the firmament or vault of the heavens, whose edges are cemented to the walls. These walls inclose the earth and all the heavenly bodies.

The whole of this theologico-scientific structure was built most carefully and, as was then thought, most scripturally. Starting with the expression applied in the ninth chapter of Hebrews to the tabernacle in the desert, Cosmas insists, with other interpreters of his time, that it gives the key to the whole construction of the world. The universe is, therefore, made on the plan of the Jewish tabernacle—boxlike and oblong. Going into details, he quotes the sublime words of Isaiah: 'It is He that sitteth upon the circle of the earth; . . . that stretcheth out the heavens like a curtain, and spreadeth them out like a tent to dwell in'; and the passage in Job which speaks of the 'pillars of heaven.' He works all this into his system, and reveals, as he thinks, treasures of science.

This vast box is divided into two compartments, one above the other. In the first of these, men live and stars move; and it extends up to the first solid vault, or firmament, above which live the angels, a main part of whose business it is to push and pull the sun and planets to and fro. Next, he takes the text, 'Let there be a firmament in the midst of the waters, and let it divide the waters from the waters,' and other texts from Genesis;

---

[1] For Eusebius, see the *Prep. Ev.*, xv, 61. For Basil, see the *Hexæmeron*, Hom. ix. For Lactantius, see his *Inst. Div.*, lib. iii, cap. 3; also, citations in Whewell, *Hist. Induct. Sciences*, London, 1857, vol. i, p. 194, and in St. Martin, *Histoire de la Géographie*, pp. 216, 217. For the views of St. John Chrysostom, Ephraem Syrus, and other great churchmen, see Kretschmer as above, chap. i.

to these he adds the text from the Psalms, 'Praise him, ye heaven of heavens, and ye waters that be above the heavens'; then casts all these growths of thought into his crucible together, and finally brings out the theory that over this first vault is a vast cistern containing 'the waters'. He then takes the expression in Genesis regarding the 'windows of heaven' and establishes a doctrine regarding the regulation of the rain, to the effect that the angels not only push and pull the heavenly bodies to light the earth, but also open and close the heavenly windows to water it.

To understand the surface of the earth, Cosmas, following the methods of interpretation which Origen and other early fathers of the Church had established, studies the table of shew-bread in the Jewish tabernacle. The surface of this table proves to him that the earth is flat, and its dimensions prove that the earth is twice as long as broad; its four corners symbolize the four seasons; the twelve loaves of bread, the twelve months; the hollow about the table proves that the ocean surrounds the earth. To account for the movement of the sun, Cosmas suggests that at the north of the earth is a great mountain, and that at night the sun is carried behind this; but some of the commentators ventured to express a doubt here: they thought that the sun was pushed into a pit at night and pulled out in the morning.

Nothing can be more touching in its simplicity than Cosmas's summing up of his great argument. He declares, 'We say therefore with Isaiah that the heaven embracing the universe is a vault, with Job that it is joined to the earth, and with Moses that the length of the earth is greater than its breadth.' The treatise closes with rapturous assertions that not only Moses and the prophets, but also angels and apostles, agree to the truth of his doctrine, and that at the last day God will condemn all who do not accept it.

Although this theory was drawn from Scripture, it was also, as we have seen, the result of an evolution of theological thought begun long before the scriptural texts on which it rested were written. It was not at all strange that Cosmas, Egyptian as he was, should have received this old Nile-born doctrine, as we see it indicated to-day in the structure of Egyptian temples, and that he should have developed it by the aid of the Jewish Scriptures; but the theological world knew nothing of this more remote evolution from pagan germs; it was received as virtually inspired, and was soon regarded as a fortress of scriptural truth. Some of the foremost men in the Church devoted themselves to buttressing it with new texts and throwing about it new outworks of theological reasoning; the great body of the faithful considered it a direct gift from the Almighty. Even in the later centuries of the Middle Ages John of San Geminiano made a desperate attempt to save it. Like Cosmas, he takes the Jewish tabernacle as his starting-point, and shows how all the newer ideas can be reconciled with the biblical accounts of its shape, dimensions, and furniture.[1]

From this old conception of the universe as a sort of house, with heaven as its upper story and the earth as its ground floor, flowed important theological ideas into heathen,

---

[1] For a notice of the views of Cosmas in connection with those of Lactantius, Augustine, St. John Chrysostom, and others, see Schoell, *Histoire de la Littérature Grecque*, vol. vii, p. 37. The main scriptural passages referred to are as follows: (1) Isaiah xl, 22; (2) Genesis i, 6; (3) Genesis vii, 11; (4) Exodus xxiv, 10; (5) Job xxvi, 11, and xxxvii, 18; (6) Psalm cxlviii, 4, and civ, 9; (7) Ezekiel i, 22–26. For Cosmas's theory, see Montfaucon, *Collectio Nova Patrum*, Paris, 1706, vol. ii, p. 188; also pp. 298, 299. The text is illustrated with engravings showing walls and solid vault (firmament), with the whole apparatus of 'fountains of the great deep,' 'windows of heaven,' angels, and the mountain behind which the sun is drawn. For reduction of one of them, see Peschel, *Geschichte der Erdkunde*, p. 98; also article *Maps*, in Knight's *Dictionary of Mechanics*, New York, 1875. For curious drawings showing Cosmas's scheme in a different way from that given by Montfaucon, see extracts from a Vatican codex of the ninth century in Garucci, *Storia de l'Arte Christiana*, vol. iii, pp. 70 *et seq.* For a good discussion of Cosmas's ideas, see Santarem, *Hist. de la Cosmographie*, vol. ii, pp. 8 *et seq.*, and for a very thorough discussion of its details, Kretschmer, as above. For still another theory, very droll, and thought out on similar principles, see Mungo Park, cited in De Morgan, *Paradoxes*, p. 309. For Cosmas's joyful summing up, see Montfaucon, *Collectio Nova Patrum*, vol. ii, p. 255. For a curious survival in the thirteenth century of the old idea of the 'waters above the heavens,' see the story in Gervase of Tilbury, how in his time some people coming out of church in England found an anchor let down by a rope out of the heavens, how there came voices from sailors above trying to loose the anchor, and, finally, how a sailor came down the rope, who, on reaching the earth, died as if drowned in water. See Gervase of Tilbury, *Otia Imperialia*, edit. Liebrecht, Hanover, 1856, Prima Decisio, cap. xiii. The work was written about 1211. For John of San Geminiano, see his *Summa de Exemplis*, lib. ix, cap. 43. For the Egyptian Trinitarian views, see Sharpe, *History of Egypt*, vol. i, pp. 94, 102.

Jewish, and Christian mythologies. Common to them all are legends regarding attempts of mortals to invade the upper apartment from the lower. Of such are the Greek legends of the Aloidae, who sought to reach heaven by piling up mountains, and were cast down; the Chaldean and Hebrew legends of the wicked who at Babel sought to build 'a tower whose top may reach heaven,' which Jehovah went down from heaven to see, and which he brought to naught by the 'confusion of tongues'; the Hindu legend of the tree which sought to grow into heaven and which Brahma blasted; and the Mexican legend of the giants who sought to reach heaven by building the Pyramid of Cholula, and who were overthrown by fire from above.

Myths having this geographical idea as their germ developed in luxuriance through thousands of years. Ascensions to heaven and descents from it, 'translations,' 'assumptions,' 'annunciations,' mortals 'caught up' into it and returning, angels flying between it and the earth, thunderbolts hurled down from it, mighty winds issuing from its corners, voices speaking from the upper floor to men on the lower, temporary openings of the floor of heaven to reveal the blessedness of the good, 'signs and wonders' hung out from it to warn the wicked, interventions of every kind—from the heathen gods coming down on every sort of errand, and Jehovah coming down to walk in Eden in the cool of the day, of St. Mark swooping down into the market-place of Venice to break the shackles of a slave—all these are but features in a vast evolution of myths arising largely from this geographical germ.

Nor did this evolution end here. Naturally, in this view of things, if heaven was a loft, hell was a cellar; and if there were ascensions into one, there were descents into the other. Hell being so near, interferences by its occupants with the dwellers of the earth just above were constant, and form a vast chapter in mediæval literature. Dante made this conception of the location of hell still more vivid, and we find some forms of it serious barriers to geographical investigation. Many a bold navigator, who was quite ready to brave pirates and tempests, trembled at the thought of tumbling with his ship into one of the openings into hell which a widespread belief placed in the Atlantic at some unknown distance from Europe. This terror among sailors was one of the main obstacles in the great voyage of Columbus. In a mediæval text-book, giving science the form of a dialogue, occur the following question and answer: 'Why is the sun so red in the evening?' 'Because he looketh down upon hell.'

But the ancient germ of scientific truth in geography—the idea of the earth's sphericity— still lived. Although the great majority of the early fathers of the Church, and especially Lactantius, had sought to crush it beneath the utterances attributed to Isaiah, David, and St. Paul, the better opinion of Eudoxus and Aristotle could not be forgotten. Clement of Alexandria and Origen had even supported it. Ambrose and Augustine had tolerated it, and, after Cosmas had held sway a hundred years, it received new life from a great churchman of southern Europe, Isidore of Seville, who, however fettered by the dominant theology in many other things, braved it in this. In the eighth century a similar declaration was made in the north of Europe by another great Church authority, Bede. Against the new life thus given to the old truth, the sacred theory struggled long and vigorously but in vain. Eminent authorities in later ages, like Albert the Great, St. Thomas Aquinas, Dante, and Vincent of Beauvais, felt obliged to accept the doctrine of the earth's sphericity, and as we approach the modern period we find its truth acknowledged by the vast majority of thinking men. The Reformation did not at first yield fully to this better theory. Luther, Melanchthon, and Calvin were very strict in their adherence to the exact letter of Scripture. Even Zwingli, broad as his views generally were, was closely bound down in this matter, and held to the opinion of the fathers that a great firmament, or floor, separated the heavens from the earth; that above it were the waters and angels, and below it the earth and man.

The main scope given to independent thought on this general subject among the Reformers was in a few minor speculations regarding the universe which encompassed Eden, the exact character of the conversation of the serpent with Eve, and the like.

In the times immediately following the Reformation matters were even worse. The interpretations of Scripture by Luther and Calvin became as sacred to their followers as the Scripture itself. When Calixt ventured, in interpreting the Psalms, to question the accepted belief that 'the waters above the heavens' were contained in a vast receptacle upheld by a solid vault, he was bitterly denounced as heretical.

In the latter part of the sixteenth century Musæus interpreted the accounts in Genesis to mean that first God made the heavens for the roof or vault, and left it there on high

swinging until three days later he put the earth under it. But the new scientific thought as
to the earth's form had gained the day. The most sturdy believers were obliged to adjust
their biblical theories to it as best they could.[1]

What further differences between White and Draper emerged from this quotation?

White's account of patristic attitudes is rather longer than Draper's. His book runs
into nearly 900 pages while Draper's was well under half that length. White is more
temperate and his Whiggishness more concealed, but he too writes of 'absurdities',
'the ancient germ of scientific truth', 'larger minded Fathers of the Church'. In
similar vein the chapter headings of his second volume are strangely reminiscent of
Picton's *The Story of Chemistry* (See page 27). They are:
13   From Miracles to Medicine.
14   From Fetich to Hygiene.
15   From 'Demoniacal Possession' to Insanity.
16   From Diabolism to Hysteria.
17   From Babel to Comparative Philology.
18   From the Dead Sea Legends to Comparative Mythology.
19   From Leviticus to Political Economy.
20   From the Divine Oracles to the Higher Criticism.

But it must be stated that White, as befits his calling, is much more scholarly in
his approach and documentation. Indeed, George Sarton regarded this as the first
scholarly book in the history of science to be produced in America. The footnotes
alone imply that. Moreover, unlike Draper, he went some way to *explain* past attitudes
and was less sweeping in his condemnation of the Church Fathers. Yet, as Dillen-
berger has observed, 'there is no reason to regard it as an acceptable scholarly book
today, much less as an adequate interpretation' (1961). Why is this? Certainly White
made mistakes—as he himself admitted—and is not an infallible guide. Nor is it
merely a question of Whiggish historiography. The fundamental flaw in the writings
of both Draper and White is pointed out in the next quotation from yet another book
of the same genre.

4.3   J. Y. Simpson's *Landmarks in the Struggle Between Science and Religion (1925)*

James Young Simpson (1873–1934) was Professor of Natural Science at New College,
Edinburgh. His father had been nephew and assistant to Sir James Young Simpson,
the pioneer in chloroform anaesthesia. In a sense he stands in the same tradition as
his two predecessors in his preoccupation with the antagonisms between science and
religion; to Draper's 'conflict' and White's 'warfare' we must now add Simpson's
'struggle'. The opening paragraph of the Preface to his book[2] speaks for itself.

> During recent years many have been visiting the battlefields of the Great War, reconstruct-
> ing in imagination, and in some instances living once again through, crises in the long
> struggle associated with this particular salient or that prominent supporting position,
> around which, with varying fortune, the tide of battle had swung from one side to the
> other. Similarly, throughout the centuries, another struggle has been engaged in by two
> opposing, though not necessarily opposed, forces, very differently equipped, and cam-
> paigning, as it seemed, on fundamentally different principles. And here, likewise, from
> the vast area of conflict stand out positions of the nature of landmarks round which the

*Figure 10   Professor
James Y. Simpson
(Photo: James Bacon and
Sons. Courtesy of New
College, Edinburgh).*

[1] For a discussion of the geographical views of Isidore and Bede, see Santarem, *Cosmographie*,
vol. i, pp. 22–24. For the gradual acceptance of the idea of the earth's sphericity after the eighth century,
see Kretschmer, pp. 51 *et seq.*, where citations from a multitude of authors are given. For the views
of the Reformers, see Zöckler, vol. i, pp. 679 and 693. For Calixt, Musæus, and others, ibid., pp.
673–677 and 761.

[2] J. Y. Simpson, *Landmarks in the Struggle between Science and Religion*, Hodder and Stoughton,
1925, p. vii.

struggle has been especially intense. The following chapters are in a sense guides to these old battlefields. In part historical, they deal with some of the most vital phases in this struggle, attempting to restate and discuss the issues, with perhaps a new ray of light illuminating the field at this point or at that.

Yet how far he has travelled from nineteenth-century dogmatism is evident from his chapter on 'early Church Fathers and science', the final part of which is reproduced below (*op. cit.*, pp. 96–7, 99–108, 111). Please read this lengthy extract in the light of the question which follows.

In two masterly volumes[1] Dr. A. D. White dealt at great length with the history of the conflict between Science and Theology, covering the ground in greater detail and with more perspective than had been traversed even earlier by the physiologist J. W. Draper.[2] No one can rise from the study of these volumes without a strange feeling of how easily a profound knowledge of the theology of the day may be accompanied by an utter lack of understanding of what Christianity is in practice, as also of the wantonness with which again and again throughout the ages the cause of Christ has been misrepresented by self-appointed agents. The whole question of the relationship of specific institutionalism to Christianity has yet to be worked out. For if history has made one thing clear, it is that in proportion to the degree in which institutionalism is regarded as primary, the spirit of Christianity departs.

At the same time, it is open to some question whether the severe strictures passed on official Christianity in these books have really been justified as a whole; or rather, whether the choice of those who have been singled out to support a general thesis to the effect that the Church through many centuries was not merely obscurantist and hostile to, but actively did its best to discourage and even suppress the investigation of natural causes, has been altogether fair. Thus, for example, of early Christian writers, Lactantius, or, as he is usually styled, Lucius Caecilius Firmianus Lactantius is again and again selected as a subject for attack.[3] As a matter of fact, singularly little is known about him with any certainty.[4] He was born in North Africa, probably shortly after the middle of the third century A.D., and is definitely known as a distinguished teacher of rhetoric in Nicomedia (Bithynia)—a post which he had to surrender on the outbreak of the persecution of Diocletian. Later he appears to have been in Gaul, in high favour with the Emperor Constantine, and entrusted by him with the education of his son Crispus. He is believed to have died at Trèves about A.D. 340. Much more of a layman than a professed theologian, he wrote with great ease, dignity, clearness, and grace, drawing largely upon his knowledge of the Greek and Roman philosophers in his defence of the Christian faith. On account of his beautiful style he was widely read, and Copernicus refers to him in the Dedication of his famous treatise to Pope Paul III.[5]

It seems, therefore, in the first place, simply a mistake to consider him in any way as representative of the recognised theological thought and attitude of mind of his day; but it is a still more serious error to judge him by isolated sentences, and especially to judge him at all without first making some endeavour to understand his general point of view, which was not so entirely unreasonable. . . .

It is further in terms of such a general background that his scornful references to such 'marvellous fictions' as the idea of the Antipodes ought to be read,[6] even if one suspects

---

[1] *A History of the Warfare of Science with Theology in Christendom*, 1896.

[2] *The Conflict between Religion and Science*, 1875.

[3] Cf. A. D. White, *op. cit.* i. 109, 209, 375, 395, etc.

[4] Cf., e.g., Introductory Notice to *The Works of Lactantius*, vol. xxi, of the Ante-Nicene Christian Library.

[5] The reference is to the *De Revolutionibus Orbium Celestium*. The passage reads: 'For it is not unknown that Lactantius, otherwise a famous writer but a poor mathematician, speaks most childishly of the shape of the Earth when he makes fun of those who said that the Earth has the form of a sphere.' Kepler also frequently quoted him.

[6] 'How is it with those who imagine that there are antipodes opposite to our footsteps? Do they say anything to the purpose? Or is there any one so senseless as to believe that there are men whose footsteps are higher than their heads? or that the things which with us are in a recumbent position, with them hang in an inverted direction? that the crops and trees grow downwards? that the rains, and snow, and hail fall upwards to the earth?' (*op. cit.* ii. 24). In chapter x. of his *Treatise on the Anger of God*, he deals in a much more effective way with the current philosophical views of the origin of the world, and the nature of affairs, and the providence of God. Cf. also *Divine Institutes*, vii. 3.

that a measure of dissatisfaction with his counter-statements is concealed in one of the concluding sentences to this section: 'But I should be able to prove by many arguments that it is impossible for the heaven to be lower than the earth, were it not that this book must now be concluded, and that some things still remain, which are more necessary for the present work.' On the other hand, there is no particular point in holding Lactantius up to ridicule[1] for his inability to accept the idea of the Antipodes, since this was an inability shared by some pagan writers as well. Naturally, however, a serious-minded man who believed that he knew where truth alone was to be found, and whose main interest was the practical one of right living in a world that might come to an end at any moment, was not going to waste his time on what he believed to be merely conjectures, and so there resulted that general haziness of idea and lack of first-hand acquaintance with the science of their day that characterise so many of the Fathers. The study of physical science was deliberately neglected and sometimes even deprecated, and the rift widened. Small wonder if it led in some cases to a certain mutual impatience with, if not disdain for, one another, on the part of ecclesiastic and man of science.

Much more outstanding and representative in this respect was Basil, Bishop of Caesarea in Cappadocia, who lived during the fourth century, having been born about A.D. 329, and died half a century later. He succeeded in combining in a very remarkable way the ideals of the monastic life with those of a great ecclesiastical administrator. His influence was largely the result of his strong personality, which gave added force to his writings. Like many of the Church Fathers, *e.g.* Origen, Hippolytus, and Augustine, he wrote a commentary upon the opening verses of Genesis, and his *Hexaemeron* is one of the best known and most admired of these treatises. The principle that he follows in this work is comparatively simple. Whenever he finds Greek science or philosophy in apparent conflict with the Scriptural narrative, he holds to the latter. At the same time, he is by no means a consistent and narrow literalist. Thus in expounding the phrase, 'And God said, Let there be light,'[2] Basil says, 'It must be well understood that when we speak of the voice, of the word, of the command of God, the divine language does not mean to us a sound which escapes from the organs of speech, a collision of air struck by the tongue; it is a simple sign of the will of God, and, if we give it the form of an order, it is only the better to impress the souls whom we instruct.'[3] On occasion he is even ready to indulge in textual criticism and emendation. Thus he gives 'Let the earth bring forth grass, the herb yielding seed after his kind' as his reading of Gen. 1 [11], adding, 'in this manner we can re-establish the order of the words, of which the construction seems faulty in the actual version, and the economy of nature will be rigorously observed.'[4]

On the other hand, he is not interested in Origen's allegorising. 'Let us understand,' he cries in another place, 'that by water is meant water; for the dividing of the waters by the firmament let us accept the reason which has been given us.'[5] Any additional details as to the characters of living things that he wishes to make use of are taken directly from the Greek or Roman men of science. He displays a remarkable acquaintance with the science and philosophy of his day, speaking of men 'who measure the distance of the stars and describe them . . . who observe with exactitude the course of the stars, their fixed places, their declensions, their return, and the time that each takes to make its revolution.'[6] Yet he is unwilling to follow their lead where it is confused. 'Those who have written about the nature of the universe,' he says, 'have discussed at length the shape of the earth. If it be spherical or cylindrical, if it resemble a disc and is equally rounded in all parts, or if it has the form of a winnowing basket and is hollow in the middle;[7] all these conjectures have been suggested by cosmographers, each one upsetting that of his predecessor. It will not lead me to give less importance to the creation of the universe, that

---

[1] Cf. J. W. Draper, *op. cit.* pp. 63, 64 and A. D. White, *op. cit.* i. 102, 103, who mentions Epicurus, Lucretius, and Plutarch (*De Facie in Orbe Lunae*, cap. vii); but there had been others, *e.g.* Herodotus (iv. 36, 42) and, in a sense, Eratosthenes and Strabo. 'The vulgar belief in a flat plain was still the vulgar or common belief when Christianity was rising to power' (C. R. Beazley, *The Dawn of Modern Geography*, i, 276).

[2] Gen. 1 [3].

[3] Hom. ii. 7. The translations are all taken from the 'Library of Nicene and Post-Nicene Fathers,' vol. viii, *St. Basil: Letters and Select Works*.

[4] Hom. v. 2; cf. also iv. 5. In the second Homily he works with the Sept. trans. of Gen. 1[2]—'The earth was invisible and unfinished.'

[5] Hom. iii. 9; cf. also ix. 1.

[6] Hom. i. 4.

[7] Plut., περὶ τῶν ἀρεσκ., iii. 10; Arist., *De Caelo*, ii. 14. References cited in trans. as above.

the servant of God, Moses, is silent as to shapes; he has not said that the earth is a hundred and eighty thousand furlongs in circumference; he has not measured into what extent of air its shadow projects itself whilst the sun revolves around it, nor stated how this shadow, casting itself upon the moon produces eclipses. He has passed over in silence, as useless, all that is unimportant for us. Shall I then prefer foolish wisdom to the oracles of the Holy Spirit? Shall I not rather exalt Him who, not wishing to fill our minds with these vanities, has regulated all the economy of Scripture in view of the edification and the making perfect of our souls?'[1] Or once again: 'Avoid the nonsense of those arrogant philosophers who do not blush to liken their soul to that of a dog; who say that they have been formerly themselves women, shrubs, fish. Have they ever been fish? I do not know; but I do not fear to affirm that in their writings they show less sense than fish.'[2] Apart from these passages, perhaps, it would be unjust to speak of Basil as showing contempt in this treatise for the Ionian philosophers. His attitude is rather one of banter; contempt is reserved for the astrologers.[3] He feels very strongly that the Scriptural account of Creation, as he understands it, is more satisfactory than any other that he knows, and he says so, with straightforward criticism. 'The philosophers of Greece,' he remarks early in the work, 'have made much ado to explain Nature, and not one of their systems has remained firm and unshaken, each being overturned by its successor. It is vain to refute them; they are sufficient in themselves to destroy one another.'[4] When he takes up a particular problem, such as 'upon what support this enormous mass (the earth) rests,'[5] it is only after an examination of various current theories, all of which seem to him insufficient, that he suggests recourse to the religious reply, 'In His hands are the ends of the earth.'[6] There is no doubt that on some points, as for example 'the essence of the heavens,' he is too quickly 'contented with what Isaiah says, for, in simple language, he gives us sufficient idea of their nature: "The heaven was made like smoke,"'[7] that is to say, He created a subtle substance, without solidity or density, from which to form the heavens.'[8] He evidently had interested audiences at his early morning addresses, who asked intelligent and difficult questions about the world of Nature,[9] dealing with points upon which Scripture provides no information, and Basil was glad to read his Aristotle—for example, the *Meteorology*, and *History of Animals*—and his Pliny, in order to be able to satisfy their curiosity. Apparently a considerable proportion of his hearers belonged to the artisan class,[10] who looked in on their way to the day's work, and it is probably on this account, as Professor Thorndike suggests,[11] that Basil sometimes speaks of God as 'the Supreme Artisan' or Artificer or Artist,[12] or bids them contemplatively 'stand around the vast and varied workshop of divine creation.'[13] 'A single plant,' he says, 'a blade of grass, is sufficient to occupy all your intelligence in the contemplation of the skill which produced it.'[14] On other occasions, he makes 'flattering allusions to arts which support life or produce enduring work, and to waterways and sea trade.'[15] . . . In breadth of outlook and genuine appreciation of scientific method, Gregory of Nyssa[16] stands far above either Lactantius or even his older brother Basil of Caesarea.[17]

[1] Hom. ix. 1.

[2] Hom. viii. 2. The references are probably to certain theories of Empedocles and Anaximander.

[3] Hom. vi. 5 ff. It was quite unnecessary for A. D. White to speak of 'the efforts of Eusebius, Basil, and Lactantius to deaden scientific thought' (*op. cit.* i. 109); they had neither the desire nor the opportunity to do so.

[4] Hom. i. 2.

[5] Hom. i. 8*b*–9.

[6] Ps. 95⁴ (Sept.). At the same time, in a later Homily (iv. 1) he refers to the 'earth, this immense mass which rests upon itself.'

[7] Is. 51⁶.

[8] Hom. i. 8.

[9] 'What trouble you have given me in my previous discourses by asking me why the earth was invisible, why all bodies are naturally endued with colour, and why all colour comes under the sense of sight' (Hom. iv. 2).

[10] Hom. iii. 1 and 10.

[11] *A History of Magic and Experimental Science*, i. 486.

[12] Hom. i, 7, 11; ii. 2; iii. 5 and 10; vi. 10.

[13] Hom. iv. 1.

[14] Hom. v. 3.

[15] L. Thorndike, *op cit.* i. 486. Cf. Hom. i. 7; iii. 5; iv. 3, 4, and 7; vi. 9; vii. 6.

[16] A.D. 335–372.

[17] Yet A. D. White makes only one direct quotation from his works (ii. 175), and J. W. Draper makes no reference to him at all.

He likewise wrote an *Hexaemeron*,[1] in which he endeavours to expound points in the history of creation in terms of the knowledge of his day, and finds that this does not conflict with his understanding of the opening passages in Genesis. He has a real conception of the Order of Nature, which he presents very clearly: 'Saying therefore that "In the beginning" the world was founded, he (*i.e.* the sacred writer) means that the occasions, causes and powers of all things were created by God, and that the essence of each of existing things came into existence at the first impulse of that will—the heavens, ether, stars, fire, air, sea, earth, animals, plants, all of which things indeed were seen by the eye of God, and shown by the reason of His power, as says the prophet, "Who sees all things before their birth."[2] Now that they had been brought together by the divine power and wisdom with a view to the perfection of each of the parts of the world, a certain necessary series followed, according to a certain order, so that any one thing first arose from the whole totality of things and appeared. Then, after that, necessarily a second followed. Then as the artificer, Nature, was necessitating, a third and a fourth and a fifth and the rest after, not by chance or fortune, as according to a certain unordered and casual movement, but as the necessary Order of Nature looks for succession in the things that arise, so, he says (philosophising in a kind of statement about physical laws), individual things came into being. He assigns even certain utterances of the ruling God to the individual things which are created, and this he does rightly and in accordance with inspiration. For whatever is done according to a certain order and wisdom of God, he refers in turn as if to a kind of definite utterance of God.'[3]

He is convinced of the Divine Reason as at work in all things. ' "Let there be light, and there was light,"[4] for speech and reason in God, as I think, are work and deed.[5] Wherefore everything that is made is made by reason. And in those things which proceed from God, nothing unreasonable and fortuitous and spontaneous is thought of. We must believe that in each actual thing there is a certain wise and productive reason, even if it surpasses our vision.'[6] He returns to this point in dealing with verse 4: 'Again, whatever is made by a certain natural order and harmony, Moses necessarily refers to the divine energy, teaching, as I think, by what has been said, that all things to be produced, following one another in a necessary order, were conceived beforehand by the wisdom of God.'[7] Or, as he phrases it in another passage: 'Truly everything that comes into being by wisdom is a word of God, not articulated by any vocal organs, but expressed by the actual wonders in things seen.'[8]

At the same time, he is somewhat disturbed by certain things in Nature. 'On this account it seems to me more reasonable that we speak on this wise. Since all things that God made are "very" good and beautiful, we ought to see the perfection of beauty in each thing. The addition of that particle "very", by means of its authoritative signification, clearly shows that there is nothing lacking to perfection. . . . For as in the animal creation we can find innumerable differences of kinds amongst them, still we agree to say by a general pronouncement that in each equally the beauty is "very" great. But this approbation is not referred to what is seen, otherwise neither the scorpion nor the land frog nor those things which are generated from rotting substances would be very beautiful. For the divine eye does not look at the appearance of things, nor is beauty defined by a certain excellence of colour or form, but by the fact that each thing in so far as it exists has a perfect nature in itself.'[9] On other occasions he can be such a literalist that in order to explain the circulation of the waters above the firmament he says that 'the back of heaven is cut up and excavated into valleys like those which are on the earth on account of the space between the mountains, that the water may be confined within them.'[10] . . .

---

[1] Unfortunately it is not translated into English, and I am indebted to Prof. H. A. A. Kennedy for much assistance in this connection.

[2] Dan. 13[42] (Sept.).

[3] Migne's edition, XLIV., Pt. i. 71 B. The trans., which is free, is based on the Greek text and parallel Latin rendering.

[4] Gen. 1[3].

[5] Τὸ ἔργον λόγος ἐστί.. The Latin rendering in Migne runs: 'oratio enim et ratio in Deo, ut ego sentio, est opus et factum.'

[6] *Op. cit.* i. 74A.

[7] *Op. cit.* i. 75 B.

[8] *Op. cit.* i. 87 C.

[9] *Op. cit.* i. 91 A.

[10] *Op. cit.* i. 90 B.

In all his references to the pagan philosophers in the *Hexaemeron*, Gregory of Nyssa is respectful; there is no suggestion of scoffing or derision. 'For I consider the firmament,' he says, 'whether it be one of four, or another beyond these, as pagan philosophers have supposed, not to be a solid hard body, but, by comparison, of an eternal and incorporeal quality which is imperceptible to touch; but the extreme of being, which is perceived by the senses, round which the nature of fire circles according to its ever moving power, is called the firmament of Scripture. . . . What indeed is heavy by nature cannot be borne aloft.'[1] Or once again: 'For those who philosophise about celestial things show that the sun itself is easily many times larger than the earth, so that not even does the shadow from the earth go forth to a great distance in the air.'[2] . . .

In Gregory of Nyssa, on the basis of his *Hexaemeron* alone, the Early Church possessed a teacher competent to take his place with the ablest of the pagan philosophers. He had a real idea of scientific method, and evidenced a worthy respect for the views of those from whom he was led to differ. Any first-hand acquaintance with his relevant writings must have led to some modification of the occasionally rather sweeping indictments by J. W. Draper and A. D. White of the Church Fathers as a whole.

What is his main criticism of the Draper/White treatment?

He chides them for generalizations based on inadequate data, citing particularly their overemphasis on Lactantius and their neglect of Gregory of Nyssa. There is also a more genuine sympathy with the particular patristic writings and certainly a reluctance to hold them up to ridicule. In many ways Simpson's work represents a half-way house between Victorian Whiggish historiography at its worst and modern history of science. Without being Whiggish about *this* we may hope that it is 'better' history than the earlier examples.

4.4   E. J. Dijksterhuis' *The Mechanization of the World Picture (1961)*

Dijksterhuis was a Professor of the Faculty of Letters and Sciences at the University of Utrecht. This notable book was first published in English translation in 1961.[3] We reprint below the whole of Section 4 of the first chapter. It is entitled 'Natural Science and Christianity', although relating only to the period of antiquity.

Figure 11 Professor E. J. Dijksterhuis (Universiteitsmuseum, Utrecht).

114. This story of the scientific thought of the Greeks has long since passed beyond the beginning of the Christian era and has thus reached the period in which by the side of pagan philosophy, which was drawing to a close and tending to decline, early Christianity was rising as a new spiritual power. It was a power of world-wide import, which claimed the complete domination of man, and which therefore interfered both with his intellectual aspirations and with his religious needs and ethical behaviour.

Thus a new factor had entered the history of thought, which involved an unprecedented restriction. In the pagan world of Antiquity, at least in the Hellenistic period, science had been able to develop practically independently of religious life. It is true that the Stoic Cleanthes gave as his opinion that the astronomer Aristarchus ought to be prosecuted for impiety ($\dot{\alpha}\sigma\dot{\epsilon}\beta\epsilon\iota\alpha$), because he contended that the earth, and consequently the central fire Hestia, were in motion,[4] but there are no signs whatever that this suggestion had any consequences, and the very fact that such an idea was mentioned and handed down as a peculiarity shows how unusual it was. Moreover, the absence of fixed, universally accepted dogma in Greek religion almost automatically ruled out conflicts with science. Science had thus always been able simply to obey its own intrinsic laws and had not been obliged to acknowledge any authority beyond that of autonomous reason. Greek thinkers had been conscious of creative mental powers, which they had unconcernedly regarded as their own property, perhaps even as their own merit.

---

[1] *Op. cit.* i. 79 C.

[2] *Op. cit.* i. 94 A; cf. also 95 D.

[3] E. J. Dijksterhuis, *The Mechanization of the World Picture* (trans. C. Dikshoorn), Oxford University Press, 1961.

[4] Plutarch (2) *c.* 6.

Entirely different views were now opposed to this: that of the helplessness of man, who by his own efforts can neither act nor think aright, but who for the former needs God's constant aid and for the latter His illumination of his mind, and that of the authority of Revelation in regard to human understanding in all provinces of life, not excluding scientific.

This altogether new conception involved a problem which was constantly to occupy and disturb men's minds and which was to lead to interminable complications and endless controversy: the problem of the mutual relations of religion and science. We shall frequently meet with its complications and pick up many an echo of the controversies. At present we are concerned with the first impact of science on Christianity as can be seen in the patristic writings.

115. Ignoring individual differences of opinion, we sum up the point of view taken by the patristic writers with regard to science as follows: more serious concerns than those of profane science are now at stake, the Christian should first of all be mindful of his salvation, and for this reason he should not desire to penetrate further into the secrets of Nature than the Scriptures demand and allow.

There are plenty of passages in which this view of scientific study is revealed: for example, St. Basil's remark that a life of Christian meekness and piety knows higher concerns and profounder solicitudes than the question whether the earth is a sphere, a cylinder, or a disc, or perhaps like a winnow, has a hole in the middle;[1] or the famous saying of Tertullian: *nobis curiositate non opus est, post Christum Jesum; nec inquisitione, post Evangelium* (for us curiosity is no longer necessary after Jesus Christ, nor inquiry after the Gospel);[2] or the warning that St. Augustine sounds in his *Confessions* against the *concupiscentia oculorum*, which is referred to in 1 John ii. 16 and which he interprets as a thirst for knowledge for its own sake.[3]

However widely this standpoint may differ not only from the views universally held in pagan Antiquity, but also from the manner in which pious scientists in later ages were to regard the relation between science and religion, it is not hard to understand why this view was originally taken. The Fathers of the Church believed that since the true salvation of mankind in Jesus Christ had now been revealed, there were for the time being no other concerns beyond helping this Gospel to find general acceptance, and that the Scriptures, in which the revelation had been given as far as possible in human language, also contained as much information about lower matters as was necessary and sufficient for man's salvation.

116. It would, however, be wrong to conclude from this that the Church Fathers were in principle antagonistic to the study of nature. To consider a thing more or less superfluous, and in certain circumstances even dangerous, is not to reject it altogether. St. Augustine urges this superfluity and this possible danger frequently and earnestly enough, but on the other hand he recognizes so wholeheartedly that worldly matters, provided they are understood aright, need not keep a man from God, who indeed manifests Himself in the world, and he emphasizes so strongly that nature, dominated as it is by numbers, also reveals eternal wisdom, that he cannot possible have condemned all science.[4]

Science, however, ought always to remain subjected to the authority of Scripture, which passes all the capacities of the human mind. St. Augustine voiced this principle emphatically, as it were on behalf of all the Church Fathers, and thus imposed on scientific inquiry a restriction which was to last for many centuries.[5]

Once this restriction had been accepted, however, science came to be regarded in quite a different light by the Fathers. In his influential *Hexaemeron*, a collection of addresses on the six days of creation, St. Basil recommends his hearers to study nature attentively, as being a work of art made by the Creator. And when after his death his friend, Gregory of Nazianze, commemorated him in an *Elogium*, he emphatically disapproved of the fact that so many Christians disparaged science, for the existence of a Creator can be inferred from the purposefulness it causes men to discern in nature; one should only beware lest one worship nature instead of Him.[6]

---

[1] Basilius, *Homilia in Hexaemeron* IX. PG. XXIX 187–8. (2) 140.

[2] Tertullian, *Liber de Praescriptione Haereticorum, c.* 7. PL II 20–21.

[3] St. Augustine (1), *Confessions* X, *c.* 35. Related views in *Enchiridion de Fide, c.* 9 and 16. PL XL 235, 238–9. *De Genesi ad Litteram* II, *c.* 9. PL XXXIV 270.

[4] St. Augustine (3), *De libero arbitrio* II, XI–XVI. Gilson, Haitjema.

[5] See, e.g., *De Genesi ad Litteram* II, *c.* 5. PL XXXIV 267. *Maior est quippe Scripturae hujus auctoritas, quam omnis humani ingenii capacitas.*

[6] Brunet et Mieli 984. Cf. Basilius (2) 5–6.

117. Although among the Church Fathers the conception that the study of nature is a Christian's duty (a conception which was to be a source of inspiration to so many pious explorers of nature) was thus already expressed, the attitude of reserve and caution predominated. In part this may naturally be explained by polemical and apologetic reasons: Greek science was bound up too closely with philosophy for it to be excepted from the general attack on the diabolical Babylon of the philosophers. Nor were its positive achievements of such a nature that it could have taken an exceptional position on that account. An even more effective influence, however, was to be exercised by the fact that the views of the most prominent patristic writers were largely determined, apart from their Christian conviction, by philosophical conceptions of Platonic or Neo-Platonic origin—two schools of thought in which the study of nature had never occupied a favourable position.

It is not hard to understand why in this period, in contrast with later ages, Platonism and its offspring Neo-Platonism were the dominant ancient influences on Christian thought. Neo-Platonism was the philosophy of the time in which several patristic writers had grown up. Origen was a pupil of Ammonius Saccas, to whom Plotinus also owed his initiation into philosophy and who is usually considered as the real founder of the Neo-Platonic school. Again, St. Augustine had been led from his original Manichaean convictions to Christianity through the study of Plotinus. Moreover, Plato's doctrine was bound to appear the most attractive of all pagan philosophies to Christians. There was a striking resemblance between the description given in the *Timaeus* of the production of the world by the Demiurge and the account of the creation in Genesis, and the suggestion that this resemblance could be explained by Plato (the Attic Moses) having become acquainted with the Septuagint in Egypt and having come into contact with the prophet Jeremiah, was readily accepted, though he was born too early for the former and too late for the latter, as St. Augustine observes.[1]

The transcendency of the realm of ideas harmonized with the Christian conception of God; the ideas themselves could be naturally interpreted as thoughts of God, of which the things on earth are imperfect realizations. Through all this a close relation between Christianity and Platonic philosophy was brought about, which was to last for as long as Augustinianism remained the predominant trend in Christian thought. No one could as yet suspect that a time was to come in which not Plato but Aristotle would furnish the philosophical basis of Christian dogma.

Thus all things combined to keep alive the sharp axiological contrast between the realms of mind and of matter, which has always been so harmful to the development of science. The thought that the whole creation to some extent participates in God's perfection and therefore deserves man's admiration and attention could not yet arise; this largely helps to explain why science was much less appreciated than might have been expected in view of the Christian conception of the genesis of the world and its lasting relation to its Creator. For although this creation has been corrupted by the Fall, so that it no longer has its pristine beauty, yet the words of St. Paul (Rom. i. 20) in which he speaks of the understanding and seeing of God's invisible things still hold good.

118. For this characteristic view of Christian thinkers we can trace the influence of the pagan philosophers. This is a first illustration of the phenomenon frequently to be observed, that in spite of its sharp opposition to Paganism in the religious and the ethical spheres, Christianity was always very ready to enjoy its philosophical treasures. For those who might think this strange or unworthy of a Christian, St. Augustine had justified such borrowings once and for all:[2] the heathens indeed possess these treasures, but they are not their rightful owners; by right they belong to the Christians, who will be the first to make the right use of them; when they appropriate these treasures, they merely do like the Jews, who, in moving out of Egypt, with the Lord's consent took with them silver and gold vessels, jewels, and precious garments of the Egyptians, because they could make better use of them. And any astonishment that so much wisdom had been conferred on peoples who had remained destitute of the light of Revelation was removed by a reference to John i. 9, where it is said that the Light lights every man that comes into the world.

119. For the proper understanding of the attitude of the Church Fathers towards Greek science it must also be noted that the latter in the third and fourth centuries A.D.

---

[1] St. Augustine (1), *De Civitate Dei* VIII, c. 11. PL XLI 235. St. Augustine leaves open the possibility that Plato might have become acquainted with the books of Moses through an interpreter.

[2] St. Augustine (1), *Confessions* VII, c. 9. *De doctrina christiana* II, c. 40. PL XXXIV 63. 'Ab ethnicis si quid bene dictum, in nostrum usum est convertendum.'

had fallen below the high level once attained by its greatest scientists. It is highly improbable that there ever was in Antiquity a universal cultural level determined by the work of authors such as Archimedes, Apollonius, Hipparchus, and Ptolemy, but it is in any case quite certain that there was no more question of this in the period now under review. The intellectual culture of those days should rather be conceived of as a curious jumble, in which relics of the flourishing period of Hellenic science were oddly blended with Oriental superstition and fantastical notions of many origins, and into which the Christian conceptions of nature now introduced a fresh element of confusion.

One may get some idea of the decline in the realm of pure science by noting the standpoint taken by different patristic writers with regard to fundamental astronomical problems. Whereas Origen, who was a contemporary of Ptolemy, appears to be quite familiar with a rather technical matter like precession,[1] Lactantius rejects even the sphericity of the earth which in the flourishing period of Greek astronomy had already been one of the elements of this science; he ridicules the notion of antipodes by making it appear as if they have their heads beneath their feet and see the heavens hanging lower than the earth[2] (in this connexion it should not be forgotten that Plutarch had made similar remarks).[3] There was a general tendency, in particular among Syrian scholars, to return to the Old Testament notions of the earth as a disc over which the sky was stretched like a tent, and when in the sixth century the geographer Cosmas Indicopleustes endorsed this in his authoritative *Topographia Christiana*, it became the prevalent opinion again for several centuries.

The work of Cosmas[4] illustrates in a general way the complicated situation that had arisen because men felt obliged to regard the statements of Scripture on scientific matters as so many divine revelations about nature, to be accepted as the literal truth. Cosmas completes the theory of the earth on a biblical foundation that had already been started by Clement of Alexandria and Theophilus of Antioch. He borrows his arguments from the account of the creation in Genesis, from Job, from Isaiah, and from the Psalms. The earth, surrounded on four sides by the Ocean, rises like a wall towards the North; the sun is carried along daily by angels round that height towards the point where it rises. The sphericity of the earth must be ruled out because the people on the other side could not then see the Lord descending through the clouds on Judgement Day.

120. Now it is indeed true that Lactantius and Cosmas must not be considered as representative of the scientific culture of the Church Fathers. A thinker such as St. Augustine does not commit such elementary errors as can be encountered in their works. Even for him, however, the difficulties with which the study of nature had to contend owing to the necessity of taking the statements of Scripture into account become quite evident. An example is his standpoint in the matter of the antipodes. Although he accepts the sphericity of the earth, he rejects the existence of people inhabiting the part of the earth turned away from us; he does not, however, do so on the strength of the primitive argument of Lactantius, but because he deems it incompatible with the unity of mankind.[5] In order to understand this argument, one has to start from the current conviction of those days that the temperate zone of the southern hemisphere was inaccessible from the North on account of the immoderate heat of the intermediate torrid zone. Any inhabitants of that zone would have to be autochthonic, and this is unacceptable because all men are descended from Adam and Eve, as stated in Scripture. Another argument followed from Romans x. 18, where it is said of the preachers of the Gospel: 'their sound went into all the earth, and their words unto the ends of the world'. As it was an established fact that they had not visited the antipodes, the latter could not exist.

121. The example is also instructive for another reason; it shows how inevitable it was for the Fathers to concern themselves with scientific questions, and how hard it was for them to persist in the standpoint that the answers to these questions did not really matter very much. Even if man's unquenchable thirst for knowledge is discounted, the interpretation of Scripture alone, which they considered one of their chief tasks and on which they all bestowed the greatest care, would have forced them again and again to concern themselves with it. The very first chapter of Genesis already gave rise to plenty of scientific questions: the reference in Gen. i. 1 to 'the heaven and the earth' created by

[1] Origenes, *Commentaria in Genesim*, PG XII 80.

[2] Lactantius, *Divinae Institutiones* III. *De falsa sapientia philosophorum*, *c*. 24. PL VI 425–6.

[3] Plutarch (2) *c*. 7.

[4] Cosmas Indicopleustes, *Topographia Christiana*. PG LXXXVIII.

[5] St. Augustine (1), *De Civitate Dei* XVI, *c*. 9. PL XLI 487. *De Genesi ad Litteram* II 9, *c*. 9. PL XXXIV 270.

God, before any separation between heaven and earth has been mentioned, the division
of the waters above and those beneath the firmament in i. 6, the creation of light in i. 3
on the first day, whereas it is not until the fourth day that the celestial luminaries appear;
these were some of the problems which the bible interpreter had to solve. It is in many
cases historically interesting to know what this solution was, because one can frequently
discern in it the echo of older Greek notions as well as the origin of familiar medieval
conceptions. It would, however, carry us too far to discuss it in more detail here: it is
enough to state that the influence of the *Timaeus* is unmistakable throughout; for their
cosmology the Fathers seem to have drawn largely from Stoic commentaries on this
dialogue, in which Aristotelian theories on the elements had also been incorporated.
For the present purpose, however, the main point is that the new source of knowledge
for scientific inquiry, which Scripture was thought to be (a view which, considering the
importance of this Book, was quite consistent), was bound greatly to intensify and confuse
the problems. If it is also remembered that science itself had long been in decadence,
it is not hard to understand that the patristic period was an era in which its revival was
not to be expected.

122. The polemical attitude often taken by the Fathers towards pagan science (when St.
Augustine recognizes the desirability of studying it, he does so with polemical intentions,
namely to enable the Christian to hold his own with pagan opponents and to protect him
against the fascination that their theories might exercise if they were new to him[1]) is most
sharply revealed in the struggle against astrology. Though genethlialogy was acceptable
to the fatalism of the Stoics, it was unacceptable to those who held the Christian conviction of man's free will to choose between good and evil. St. Augustine therefore—to
take just one example—enumerates all the traditional arguments already found in
Carneades, and he appears to attach great value in particular to the twin-argument,
for by this even the moderate view that the stars function as omens and not as causes is,
in his opinion, refuted.[2]

123. However uncompromising and unyielding his rejection of astrology may seem to
be, upon closer consideration it has some weak points, which were not insignificant in
the further development of the relations between Christianity and astrology. In the first
place his condemnation does not extend to the doctrine of the influence of the stars on
purely physical properties; no more than any other thinker of Antiquity could St. Augustine help being impressed by the evident interdependence of natural phenomena on earth
and the positions and movements of certain heavenly bodies (the alternation of the
seasons in connexion with the sun, the influence of the moon on the tides), and by the old
popular belief in the connexion between processes of growth on earth and the different
phases of the moon. In this way the door had already been opened, at least in principle,
to the recognition of the value of astrology in medical science. Subsequently, however,
just like Origen[3] and Lactantius[4], he clearly intimates that he looks upon the ambition
of the astrologers to read men's fortunes from the stars as a sinful rather than a futile
pretension. It is not at all consistent with his views to reject a *praescientia futurorum*, a
foreknowledge of what the future has in store, as impossible; he disapproves strongly of
the fact that Cicero does, and he calls it the most patent madness to believe in a God and
deny that He at any rate knows the future.[5]

But man should not presume to penetrate into the divine mystery, and that is the very
thing the astrologer attempts to do. What is sinful in this? The answer is as surprising
as it is significant: he can only succeed with the aid of demons, i.e. of God's enemies.
Thus the permissibility of astrology has indeed been disposed of, but not its possibility.
For if the demons are able to read the future from the stars and to allow the astrologer
who has called in their aid to participate in their knowledge, it follows that the future
must be written in the stars in some way or other, for not even demons can read something
where there is nothing. It is obvious that the matter could not be considered as settled,
and indeed the standpoint that a Christian ought to take with regard to astrology was to
be repeatedly discussed and qualified.

---

[1] St. Augustine (1), *De Genesi ad Litteram* I, *c*. 19, II, *c*. 1. PL XXXIV 261, 264.

[2] St. Augustine (1), *De Civitate Dei* V, *c*. 2–7. PL XLI 142–8.

[3] Origenes, *Commentaria in Matthaeum* XIII, *c*. 6. PG XIII 1107–8.

[4] Lactantius, *Divinae Institutiones* II. *De Origine erroris*, *c*. 17. PL VI 336.

[5] St. Augustine (1), *De Civitate Dei* V, *c*. 9. PL XLI 149.

124. The Fathers not only admit the possiblity of practising efficient astrology with the aid of demons, they also accept the reality of a magic art practised with the same aid. But for this very reason the Christian ought to abstain from it. With the advent of Christ the contest with Satan and his hosts has begun; the gentiles were still at liberty to avail themselves of their aid but the Christians are no longer allowed to do so. In the person of the three Magi from the East—that is the interpretation of Tertullian[1]—the magic art came to surrender its dominion at the manger of the Saviour, and this abdication is expressed by the fact that they returned to their country by a different route from that by which they had come (Matt. ii. 12).

What characteristics of historical writing, absent in the earlier quotations, are shown by this extract?

This is a much more recent example of historical writing and shows many of the features that would today be classified as virtues. Note first how Dijksterhuis handles his primary source material and what kinds of inferences he allows himself to make. A typical case is his evaluation of the comment by Cleanthes. Instead of generalizing on the basis of this remark he does the opposite, asserting that the opinion voiced must have been sufficiently rare and untypical for it to have been thought worth recording. This, of course, is borne out by a wide acquaintance with the writings of the period, which also enables him to note, as Simpson did, that Lactantius and Cosmas were not truly representative.

Then it is also apparent that the author really has tried to get inside the mental attitude of those early writers. That is why he so freely quotes Scriptures, not for reasons of piety but to help us understand the past. This unwhiggish approach is also demonstrated in his failure to ridicule such early pseudo-sciences as astrology. Scientifically they are no longer of any value, but they played a prominent part in the thought and even the science of an earlier age.

There is a third way in which this extract has advanced significantly beyond the one before it. Dijksterhuis places far more stress on the Greek intellectual heritage received by the Church Fathers. He is not merely content to observe their reaction to Biblical concepts. This recognition of a dual 'input' into early scientific thinking is crucially important, and will play an important part in the present course.

Finally, note how Dijksterhuis handles the 'warfare model' of which we spoke in the Introduction. He does not abandon it but presents it in a very modified form. Study of theology is seen as inimical to the study of nature, but mainly in terms of conflicting priorities on a man's time, not as representing a fundamental ideological antithesis. How far it represents an acceptable view of history is for you to decide.

---

[1] Tertullian, *De Idolatria*, c. 9. PL I 672.

Further Reading

A. R. Hall, 'Can the History of Science be History?', *Brit. J. Hist. Sci.*, 1969, **4**, 207–220.

A. G. Debus, 'The History of Chemistry and the History of Science', *Ambix*, 1971, **18**, 169–177.

T. S. Kuhn, 'The Relations between History and the History of Science', *Daedalus*, Spring 1971, pp. 271–304.

Acknowledgements

Grateful acknowledgement is made to the following sources for material used in this unit:

Text

G. Bell & Sons Ltd. for H. Butterfield, *Origins of Modern Science;* The Clarendon Press for E. J. Dijksterhuis, *The Mechanization of the World Picture* (trans. C. Dikshoorn) © 1961 Oxford University Press. Reprinted by permission of The Clarendon Press, Oxford; Hodder and Stoughton Limited and Mrs. H. D. Simpson for J. Y. Simpson, *Landmarks in the Struggle between Science and Religion*, used by permission of the publishers, Hodder & Stoughton. Copyright.

Illustrations

Figures 1 and 6: Science Museum (by permission of the Controller, HMSO); *Figure 3:* Mansell Collection; *Figure 4:* Uitgeverij 'De Tempel' for *Osiris; Figure 5:* Smithsonian Institution, Washington for ISIS; *Figure 7:* American Chemical Society: *Figure 8:* Mansell Collection (Uffizi Photo); *Figure 9:* Cornell University Archives; *Figure 10:* New College, Edinburgh (Photo: James Bacon and Sons); *Figure 11:* Universiteits-museum, Utrecht.

The Impact of the Copernican Transformation

Prepared by Professor R. Hooykaas for the Course Team

Contents

1 Introduction

This unit is closely related to the first television programme which is specially important if you are not familiar with the background to Copernicus' great achievement. It illustrates certain of the astronomical problems involved in a way that would be impossible in print. The first radio broadcast deals with some other important intellectual issues underlying the Copernican 'revolution', and the second one with an important aspect of its aftermath. The set book *Religion and the Rise of Modern Science*[1] will be found essential for supplementing some of the points in the text of this unit.

General Aim

To explore the impact of Copernicus' theory on contemporary thought and its relationship to religious thinking.

Objectives

After studying this unit, the set book and television and radio programmes, you should be able to:

1 Identify the main features of the Graeco-Roman and Biblical world views and demonstrate their influence on medieval thought.
2 State the main features of Greek astronomical theory.
3 Discriminate between *mathematical* (or *astronomical*) and *physical* models of the universe.
4 Outline Copernicus' theory.
5 Evaluate the influence of antiquity on Copernicus.
6 Describe and evaluate the attitudes to Copernicanism of contemporary and modern writers.
7 Explain how the Biblical approach was related to Copernicanism.
8 Define the principle of *accommodation* and its use by Calvin.
9 Identify the common elements in Copernicanism and the theories of Tycho Brahe.
10 Outline the reception given to Copernicus' theory.
11 Describe the theological problem of the plurality of worlds.
12 Identify ancient attitudes to the problem.
13 Evaluate the effects of the spread of the idea of plurality of worlds, its accommodation to the Biblical world view and its relationship to the rise of mechanistic philosophy.

---

[1] R. Hooykaas, *Religion and the Rise of Modern Science*, Scottish Academic Press/Chatto and Windus (SET BOOK).

2 The Twofold Heritage from Antiquity

The Western intellectual heritage goes back to two sources, the Graeco-Roman and the Hebrew traditions. Few people nowadays are versed in knowledge of the Greek and Roman writers and, in general, we are less familiar with the Bible than our grandparents were. The more, however, one knows about classical culture and the contents of the Bible, the more one will discover that Athens and Jerusalem are omnipresent in our literature (Shakespeare, Milton), painting (Rembrandt), law, philosophy and, though less conspicuously, even in our science. Consequently, the decrease of direct knowledge of these two sources of our civilization might lead to serious deficiencies in our understanding of the history of Western culture.

At first sight it may seem strange that a course on the history of *science* deals with these two trends of thought. Science, as we are prone to think, tries to describe and explain nature as it really is, and it has nothing to do with Greek philosophy and Biblical religion. Yet, though we may try hard to effect a methodological separation of our science from aesthetic, philosophical and religious viewpoints, we cannot effect a psychological separation. The human mind is *one*, however much we try to keep our science of nature free from theological, philosophical, social, and other non-scientific preconceptions and prejudices. There is nothing against this, as long as we are aware of it; yet, without some 'belief' (e.g. the belief that there is order in nature), there would hardly be any stimulus to scientific research; and again without some 'belief' (e.g. that it is beneficial to mankind, or that it is part of what Boyle called 'rational worship') it would be to no purpose.

Reflection on our own scientific activities may make us more conscious of what it is all about; why we cultivate science, how we do it, what are its limits and limitations.

When doing this as historians, then, we will meet again and again with these two roots of our intellectual culture: the Graeco-Roman and the Biblical world view. Now these two show fundamental differences in their approach to religious, ethical, philosophical and scientific problems. Though the Church Fathers and the medieval and Renaissance philosophers tried to weld them together into a unity, the result suffered from inner tensions and a continual struggle for hegemony of the one over the other. The Greek philosophers, in particular Aristotle, acquired their authority over the medieval mind with great difficulty, and there always remained uneasy feelings which led again and again to a revolt. Nevertheless, Greek philosophy and astronomy gave our ancestors a world picture that proved itself satisfactory up until the sixteenth century. The Biblical authors, on the other hand, seemed to have a rather naïve world picture. In general, they described the world as we see it: a flat earth, under the blue sky across which the sun, the moon and the stars are moving. When a more sophisticated conception comes in the Bible, this is not conformable with the Aristotelian world picture; thus the Book of Genesis speaks about 'the waters above the heaven', whereas in Aristotelian physics water can only occur below the heaven. That is, the rather 'naïve' world picture of the Bible and the more sophisticated scientific world picture of the Greeks are incompatible, and here, too, the Church Fathers had to make great efforts to reach a degree of reconciliation between the two. The matter became more complicated because there were considerable differences within the Greek tradition: Copernicus, though remaining to a certain extent within the Aristotelian tradition, deviated from it by following another Greek tradition also.

One important point in our course will be to see how far the opposition to the 'new' world picture stems from Biblical (or allegedly Biblical) sources, and how far it finds its roots in alternative Greek ways of thought. We will have to take into account that the classical and the Hebrew heritages had so strongly merged, that people often believed they were maintaining a Biblical position when they were just reading their Bible through the spectacles of Greek philosophy and cosmology.

Much more fundamental than the problems caused by the world picture are those caused by the world view. By the term *world picture* we understand the physical model of the structure of the world (i.e. that of a central earth, surrounded by con-centric spheres bearing the planets). By *world view* we mean an evaluation of the world (e.g. whether the world is divine or not, alive or not, whether or not it is liable to change, and whether eventual change ensues from an immanent cause or from without).

It is evident that the Biblical authors hold a world view that is to a large extent incompatible with the Greek one, e.g. in that the former consider the world as having been made by a Divine Creator who infinitely transcends human reason, and as dependent on His continual sustenance, whereas the latter considers Nature as a self-sufficient and self-supporting divinity (see *Hooykaas* (SET BOOK), chapter 1). This leads to different trends in scientific method (*op. cit.*, chapter 2) and to different conceptions of scientific aims (*op. cit.*, chapter 3).

Matters become more complicated because it is impossible to give a hard and fast rule about which part of the Greek heritage was acceptable from a Christian view-point. A total rejection was very rare and even then it often meant that a non-Christian tradition was mistakenly put forward as essentially Biblical (e.g. Paracelsism and Hermetism). The founders of modern science in the sixteenth and seventeenth centuries recognized the necessity of reviving the best of the Greek scientific tradition, and at the same time they gradually recognized that essential corrections of the data, doctrines and methods of Greek science were indispensable for further progress. It dawned upon them that not only the world picture but also the world view needed correction. It should be stressed, however, that by no means everything is explained by revealing the influence of 'beliefs'. There are also the geographical, social, political and economic factors, and the great difficulty for the historian is that material factors are so closely intertwined with spiritual ones that in many cases it is well nigh impossible to ascertain which of the two are the causes and which the effects (cf. the problem of Puritans and Science). The study of our subject will show that beliefs (even 'wrong' beliefs) may act as a stimulus to research but that they may also stifle further investigation of nature. It will demonstrate, also, that scientists, though trying not to let human prejudice interfere with their results, are not so impassive as might appear from scientific textbooks.

Having stressed the dual nature of the intellectual heritage at the time of the Renaissance we shall now look at two main aspects of the new Copernican science.

Figure 1 The pre-Copernican universe named after the Greek astronomer Ptolemy, from an original engraving. The sublunar world contains the four elements earth, air, water and fire (Mansell Collection).

3 The Motion of the Earth

3.1 Copernicus' system

The Copernican 'Revolution' was no revolution. As we now see, it may have *inaugurated* a revolution in science, but this took one hundred and fifty years to accomplish itself and before 1600 it went on at a very slow pace. Even then the acceptance of the Copernican system was far from general. One historian of astronomy even wondered whether there *were* Copernicans before 1600. Of course, this is a bit of an exaggeration. Copernicus had one pupil in the direct sense, namely Joachim Rheticus, a professor of Wittenberg, who stayed with him for almost two years and who wrote the *First Narrative* of Copernicus' work (printed in Danzig, 1541). In England Thomas Digges (in 1573 and again in 1576) declared his adherence to the Copernican system and there were a few other believers in Copernicus' doctrine of the motion of the earth. But, though in the second half of the sixteenth century Copernicus was generally recognized and admired as a great astronomer, this did not imply acceptance of his thesis that the sun stands still in the centre of the universe and the earth revolves around it in a period of one year and, moreover, revolves around its own axis in twenty-four hours, as being physically or objectively true. The publication of Copernicus' work in 1543 did not cause the stir we connect with the concept of revolutions.

The Portuguese discovery that the tropics are habitable, which had become generally known half a century earlier, sent a wave of wonder throughout Europe, and the rise of a new star in 1572 caused not only great surprise but even great anxiety. In comparison with these disturbances the publication of Copernicus' hypothesis was just a ripple on the surface. Why then this easy acceptance of Copernicus as a professional astronomer and this indifference or even opposition to his theory of the motion of the earth?

In order to answer this question we have to consider first the self-imposed task of ancient astronomy, namely the reduction of all apparently irregular movements of heavenly bodies to perfectly circular and uniform motions. It was considered impossible that divine and perfect bodies would perform irregular, non-uniform, that is 'imperfect', motions. Appearances of irregular planetary motions were plainly seen by the eye, and yet this irregularity could not be true. Consequently, *phenomena* ought to be *saved* by demonstrating that they may be derived from a combination of regular circular motions. The first solution of this problem was given by Plato's friend Eudoxus, whose system was adopted with some modifications by Aristotle. The universe is a globe, the centre of which is occupied by the earth. Inside this globe are several concentric spheres. The smallest one bears the moon, next follow the spheres of the planets Mercury and Venus, that of the sun and those of the planets Mars, Jupiter and Saturn, whereas the fixed stars are on the outmost sphere which turns round in one day and night (Figure 1).

The five planets, however, are seen to have irregular movements: they progress, stand still and make a retrograde movement and then progress again. In order to explain these irregularities Eudoxus let each planetary sphere partake in several rotations (around different axes) at the same time. A weak point in the theory is that the diameter of a planet would always appear to be the same as its distance to the earth would be unchangeable. This is contradicted by observation.

3.2 Mathematical hypotheses

In an attempt to overcome this difficulty (and other ones) later Greek astronomers resorted to other devices: the excentric, the epicycle and the equant. In the following diagrams E represents the earth, P a planet, S the sun, C the centre of a circle.

3.2.1 The excentric

A simplified example of the application of an excentric is provided by the motion of the sun: the sun is said to move uniformly in a circle, though the earth is not in in the centre C but displaced some way from it. The angular velocity with respect to the centre is uniform, but with respect to the earth it is not.

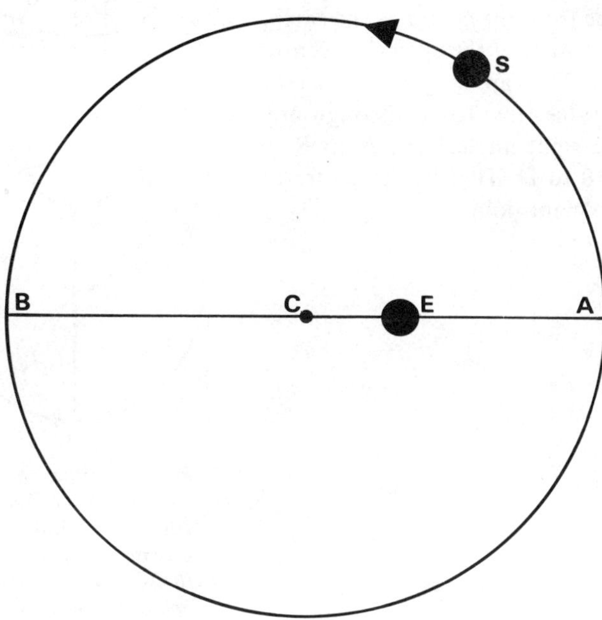

3.2.2 The epicycle

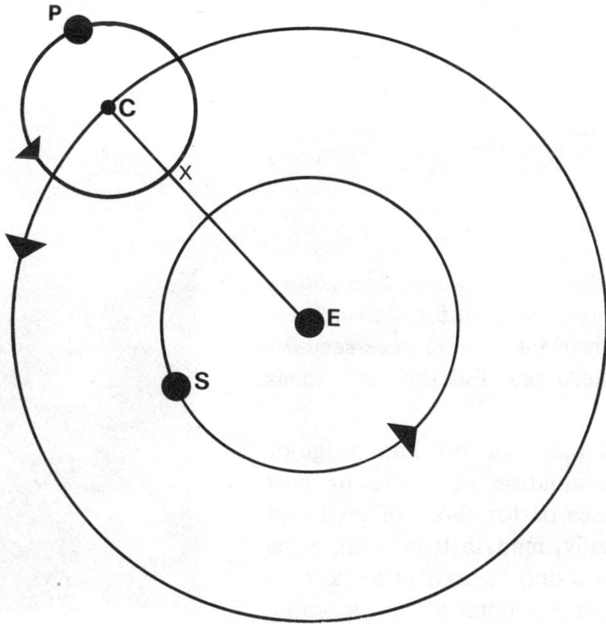

This device involves a combination of two circular motions. A planet is thought of as moving on a circle (epicycle) the centre of which (C) shares the rotation of a greater ('leading') circle (deferent) around the earth. In this way loops of retrogression could be accounted for since, at X, the planet does actually seem to be going backwards.

Ptolemy used epicycles and excentrics in many kinds of combinations to account for the paths of the planets as they can be seen by a terrestrial observer.

3.2.3 The equant

Ptolemy (fl. 150 A.D.) introduced a device—the equant—which was not conformable to the rule that only *uniform* circular motions should be admitted in astronomical theories. (A uniform circular motion has a constant angular velocity with regard to the centre of the circle.)

Imagine the planet P moving in an epicycle as before, and the earth E in an excentric position. Let the epicycle's centre rotate with the deferent which has the centre C. The equant point (Eq) may be taken at the same distance from the centre as the earth, but on the opposite side. It was now supposed that the motion of the epicycle's centre would be uniform only if viewed from the equality point. Thus an imaginary observer there would see the epicycle's centre always take the same time to go through any given angle. For instance, it would move through the right angle from A to B in exactly the same time as through the right angle from B to D. That is, the centre of the epicycle has a uniform angular velocity about the equant point.

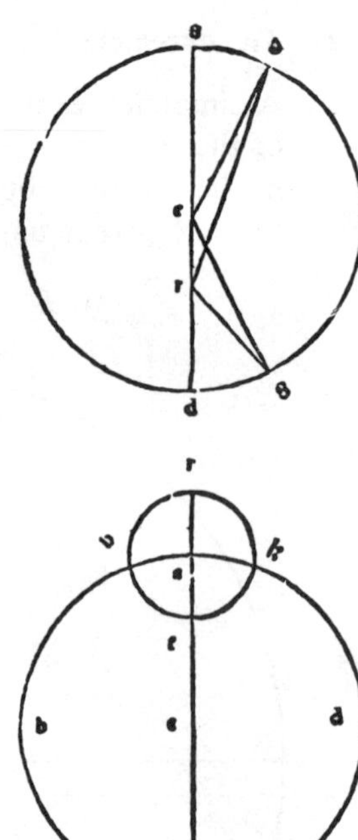

Figure 2 *Diagrams from 1515 edition of* The Almagest *showing* (top) *eccentric circle and* (bottom) *deferent and epicycle (Ronan Picture Library and Royal Astronomical Society).*

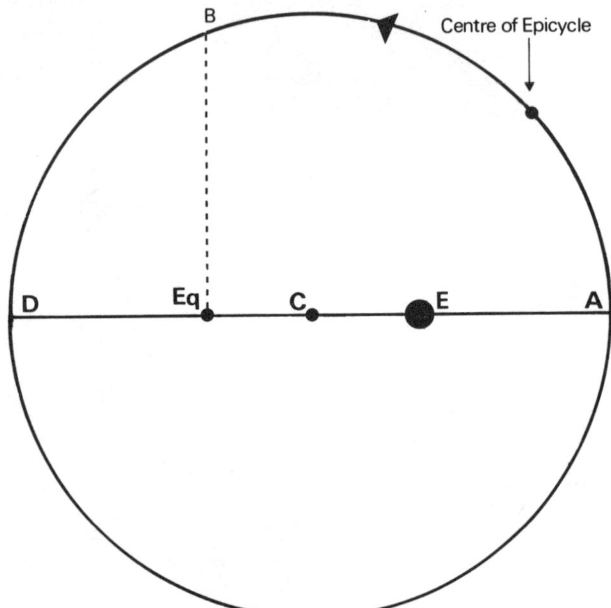

In this way Ptolemy treated each of the planets separately and obtained satisfactory agreement with the results of observation. It now became the task for astronomers up till Copernicus' time to introduce such minor modifications as were necessary for the Ptolemaic system to be given an ever-increasing accuracy. But the basic ideas remained the same throughout that long period.

Astronomy was one of the liberal arts; that is, its first aim was not knowledge or science but practical achievement, in this case determination of future or past movements of heavenly bodies for astrological purposes or for those of civil and religious life, e.g. fixing the dates of festivities. Gradually, many astronomers came to hold that their theories, then, need not be *true*; they had only to serve as a scaffolding, as a means to arrive at a desired practical end. From that point of view it would not be so important whether the hypothetical motions were strictly uniform or not, and even whether the earth has a central position and is standing still or not. The only condition they had to fulfil was their being easily accessible to computation. Yet, it seems that even with these 'mathematical' hypotheses ('mathematical', not so much because mathematics played a large role in them, but rather because they were used by 'mathematicians', that is astronomers) there were some *physical* conditions, namely those of the central position of the earth and uniformity and circularity of motion, which had to be fulfilled.

Figure 3 *Supposed portrait of Ptolemy, engraved by G. Cook (from a print in the Royal College of Physicians).*

Astronomical theories like that of Eudoxus, which pretended to be objectively true, were called in a later epoch 'physical hypotheses', because they were used by *philosophers*, people who contemplated the true being, the nature or *physis* of things. The *art* of astronomy, then, created 'mathematical' hypotheses, fictions to help in computing the courses of the heavenly bodies; the science or philosophy of physics, on the other hand, put forward *physical* hypotheses, claiming to give a faithful image of the universe, founded upon knowledge of the intrinsic nature of things.

1 What was the common basis of Greek astronomical theories?
2 What major difficulty did the Greeks face in their systems?
3 What was the basic difference between 'mathematical' and 'physical' hypotheses?

1 That the heavenly bodies moved in uniform circular motion.
2 The apparent movement of some heavenly bodies (the planets) is not regular.
3 'Mathematical' hypotheses did not have to be correct, they only had to provide models from which by mathematical calculation the place of the planets could be found. 'Physical' explanations attempted to show objective truth; to provide a *physical* model, an attempt to establish a *comprehensively correct* relationship between *all* the relevant phenomena.

3.3 Astronomy a part of physics

Now some astronomers were not contented with the position of astronomy as a kind of applied mathematics, but they considered astronomy as a part of physics or as revealing the 'mathematical aspect of physical things'. The editor of Copernicus' work, Andreas Osiander (1498–1552), added a Preface of his own *To the Reader: concerning the hypotheses of this work*, part of which is reproduced below.

Figure 4 *Andreas Osiander, 1498–1552 (Staatsbibliothek, Berlin: Preussischer Kulturbesitz).*

CONCERNING THE HYPOTHESES OF THIS WORK

The author of this work has done nothing blameworthy, for it is the duty of an astronomer to compose the history of the celestial motions through careful and skilful observation. Then . . . he must conceive and devise, since he cannot in any way attain to the true causes, such hypotheses as, being assumed, enable the motions to be calculated correctly from the principles of geometry, for the future as well as for the past. The present author has performed both these duties excellently.

It is quite clear that the causes of the apparent unequal motions are completely and simply unknown to this art. And if any causes are derived by the imagination, as indeed very many are, they are not put forward to convince anyone that they are true, but *merely to provide* a correct basis for calculation. Now when from time to time there are offered for one and the same motion different hypotheses . . . the astronomer will accept above all others the one which is easiest to grasp. The philosopher will perhaps seek the semblance of the truth. But neither of them will understand or state anything certain, unless it has been divinely revealed to him.

Let us therefore permit these new hypotheses to become known together with the ancient hypotheses, which are no more probable; let us do so especially because the new hypotheses are admirable and also simple, and bring with them a huge treasure of very skilful observations. So far as hypotheses are concerned, let no one expect anything certain from astronomy, which cannot furnish it, lest he accepts as the truth ideas conceived for another purpose, and depart from the study a greater fool than when he entered it. Farewell.

In the printed version this Preface was followed by a *Preface of the Author*, which began as follows:

TO THE MOST HOLY LORD, POPE PAUL III. THE PREFACE OF
NICOLAUS COPERNICUS TO THE BOOKS OF THE REVOLUTIONS

I may well presume, most Holy Father, that certain people, as soon as they hear that in
this book *On the Revolutions of the Spheres of the Universe* I ascribe movement to the
earthly globe, will cry out that, holding such views, I should at once be hissed off the
stage. For I am not so pleased with my own work that I should fail duly to weigh the
judgment which others may pass thereon; and though I know that the speculations of a
philosopher are far removed from the judgment of the multitude—for his aim is to seek
truth in all things as far as God has permitted human reason so to do—yet I hold that
opinions which are quite erroneous should be avoided. . . .

That I allow the publication of these my studies may surprise your Holiness the less in
that, having been at such travail to attain them, I had already not scrupled to commit to
writing my thoughts upon the motion of the Earth. How I came to dare to conceive such
motion of the Earth, contrary to the received opinion of the Mathematicians and indeed
contrary to the impression of the senses, is what your Holiness will rather expect.

Was Copernicus' attitude to hypotheses, as revealed in this extract, different from Osiander's?

Copernicus seems to be looking for the *true* explanation—which would involve a
physical model, conformable to reality.

Some readers may have been misled by the first Preface though a careful reader
could see that it was not by Copernicus himself.

1   It is followed by the 'Preface of the *Author*', under the heading 'The Preface of
    Nicholas Copernicus to the Books on the Revolutions' thus implying a contra-
    distinction.
2   It considers the astronomical hypothesis as a *mathematical* one, whereas
    Copernicus' own text clearly puts forward the *physical* character of his system of
    the world.
3   It speaks about the author of the book in the third person, whereas the author's
    Preface speaks about him in the first person, and mentions the author's name.

3.4   The antiquity of Copernicus' theory

Copernicus fully shared the conviction that all movements in the heavens must be
circular and uniform: his studies deal with the 'divine circular movements of the
world'. The question why Copernicus' work was so enthusiastically welcomed by
some people who could hardly be called 'Copernicans' finds its answer in the fact that
Copernicus removed the equant from astronomy. It is probable that this offence
against the uniformity of heavenly motions, which had been rather easily overlooked
by the astronomers of the old school, gave rise to his dissatisfaction with the Ptolemaic
system. As Copernicus put it:

> But even if those who have thought up excentric circles seem to have been able for the
> most to compute the apparent movements numerically by those means, they have in the
> meanwhile admitted a great deal which seems to contradict the first principles of regularity
> of movement.

*Figure 5   Nicolaus
Copernicus, 1473–1543,
copy at Strasbourg
(Ronan Picture Library).*

In his opinion, the assumption of irregular motions would imply inconstancy in the
nature of the moving cause, or irregularity in the moved body. But 'the mind shudders
at such suppositions; it is quite unfitting to suppose that such a state of affairs exists
among things which are established in the best system', regular movements only
'appear to us as irregular'. Therefore, Copernicus constructed a system without
equants. It was considered a great feat of 'mathematical' skill that in this way he
restored astronomy to the purity of exclusively uniform motions. Nicholas Mulerius,
who edited the third edition of this work, in spite of his being a non-Copernican
himself, gave it the title *Restored Astronomy*.

On the other hand, Copernicus clearly foresaw the opposition to his work:

> The scorn which I had to fear on account of the novelty and absurdity of my opinion almost drove me to abandon a work already undertaken.

As to the novelty he could immediately reassure his readers. He recognized that he 'expounded many things differently from my predecessors'. But this 'newness' (of which he was proud) basically was a *return* to unspoilt ancient wisdom. Copernicus was one of the typically humanistic Renaissance philosophers, and to them 'novelty' mainly meant newness in relation to the immediate past, though in fact signifying a return to the most remote past. It was a *restoration*, not a revolution that they had in mind. Their aim was conservative and not progressive; truth had to be rediscovered. The religious Reformation, too, was meant to be a restoration (a re-formation) of Catholic Christianity; the same accusations of novelty were levelled against it by its opponents and it made the same claims to restoring the truly ancient and authentic doctrine in its original purity.

Copernicus became dissatisfied with the astronomy of the medieval schools because they had *different* mathematical solutions for the same astronomical problem, instead of the *one* true system conformable to the real world, and because they had abandoned uniformity. His first move was:

> To re-read all the books by philosophers which I could get hold of to see if any of them even supposed that the movements of the spheres of the world were different from those laid down by those who taught mathematics in the schools.

And so he found that some Pythagoreans and Heraclides had let the earth move, and he found also (though it was omitted from the published version) that Aristarchus of Samos let the earth revolve on its axis in one day and turn round the sun in one year.

The fact that he was able to refer back to these ancient authorities to support his theories meant that Copernicus could not be accused of being a lover of novelties. His supporters and most of his opponents recognized this by dubbing his system the Pythagorean, or Copernico-Pythagorean, or Aristarchian theory.

Why was Copernicus' theory a new beginning in spite of his going back to Antiquity? (Note particularly Sections 2, 3.3 and 3.4.)

Although there were ancient astronomers from whom Copernicus was able to derive support for his views, those were in contrast to the Ptolemaic thesis, which was almost universally accepted by contemporary astronomers. He was concerned to rediscover the genuine nature of the cosmos, not merely to provide a new basis for mathematical calculations. We may also note that Copernicus obtained one connected and ordered system to replace the independent systems for each individual planet of Ptolemaic astronomy. (See also Section 3.6.)

3.5 The absurdity of Copernicus' theory

3.5.1 Physical arguments

Copernicus recognized that his idea that the earth rotates is 'almost in opposition to common sense'. Some of the arguments against it are mentioned by Copernicus himself. Loose objects would be flung away from the surface of the earth, clouds would always seem to be floating to the west, and bodies freely falling from a tower would arrive to the west of the base of the tower.

'Absurdity' existed also with respect to the tenets of the Aristotelian natural philosophy inculcated in the minds of people for many centuries, so much so that they had become part of 'common sense'. For instance Aristotle's heaviest element, earth, is borne by a natural tendency towards the centre of the universe along a rectilinear path. Being an element it can only have one natural motion and for terrestrial elements this is the rectilinear one. If it had a circular motion, this would be unnatural and it would not last long. Only heavenly bodies consisting of their own special element (aether) could have by nature a perpetual circular motion.

Copernicus, however, had his physical counter-arguments and his friends persuaded him that 'the fog of absurdity would be dissipated by my luminous demonstrations'. When his opponents say: the earth is not a heavenly body, and, consequently has to be in the centre of the universe which is its natural place, Copernicus answers: each body tries to unite with the whole to which it belongs. Bodies of the same nature attract each other, earth seeks earth (and not the centre of the universe). When the opponents say that earth, being a heavy element, cannot have a circular but only a rectilinear motion, Copernicus answers that the earth is spherical and that a rotatory motion therefore is its natural motion (i.e. the cause of rotation was geometric form). In these two answers Copernicus implicitly denies the Aristotelian dogma that there is a fundamental difference between heavenly and terrestrial bodies.

According to the Aristotelian view which prevailed in the Middle Ages and gradually succumbed under the attacks of New Philosophy in the sixteenth–seventeenth centuries, all terrestrial substances, made of the four elements earth, water, air and fire, are corruptible. They have by nature only rectilinear motions, either towards the centre of the universe (as in the case of earth and water) or away from it (as air and fire). The heavenly bodies, however, consist of a fifth element which by nature has a circular motion and which is incorruptible. The heavenly and earthly regions were demarcated by the sphere of the moon; consequently, whereas in the sublunar world no stability reigns, in the superlunar world there is no change at all. Copernicus' opinion, however, implies that matter of the moon, when separated from its main body, would show gravity just as earthy matter, when separated from the main body of the earth, tends to reunite with it. Conversely, the earth has as its natural motion a circular movement, just like the so-called heavenly bodies. These tenets are indissolubly connected with the conviction that the earth itself is one of the 'heavenly' bodies.

3.5.2 Metaphysical arguments

Copernicus does not use only astronomical or mathematical and physical arguments for the motion of the earth. He often refers also to the beauty, harmony and divine order of the cosmos. The universe is spherical 'because this figure is the most perfect of all'. Falling bodies try to reunite with the body to which they belong, for 'nothing is more repugnant to the order of the whole and to the form of the world than for anything to be outside of its place'.

Aristotelians considered the state of immobility as more noble and divine than that of change and instability. Copernicus, however, deems it absurd to ascribe movements to the immense sphere of the fixed stars which contains all things, rather than to one of the bodies contained in it, namely the earth. As to the *annual* motion of the earth, which had hardly been taken into consideration by the Aristotelians, Copernicus connects it with the revolution of the other planets around the sun, and in this way he pretends to avoid disturbing human understanding 'by an almost infinite multitude of spheres, as those who keep the earth at the centre of the world are forced to do'. But Copernicus wants 'rather to follow the wisdom of nature, which as it takes very great care not to produce anything superfluous or useless, often prefers to endow one cause with many effects'. The *one* annual movement of the earth round the sun explains the progressions and retrogradations of the five planets, and the one daily rotation explains the apparent rotations of all heavenly bodies in that time:

> In the midst of all dwells the sun, for who would place this lamp of a very beautiful temple in another or better place than that from which it can illuminate everything at the same time? . . . And so the sun, as if resting on his royal throne, governs the family of the stars which circle around.

Copernicus then points out that in his system there is a 'wonderful symmetry and a sure bond of harmony for the movement and magnitude of the orbits'; and this he illustrates by some phenomena which seem rather haphazard in the old system whereas they necessarily follow from his own suppositions. He becomes so enthusiastic about his discovery that he exclaims: 'so great is the divine handiwork of the Most Excellent and Greatest One'. His exposition reveals that he, like Kepler after him, was strongly influenced by Platonism and Pythagoreanism as revived in the fifteenth century in Italy.

It should be pointed out, however, that the description of his system which Copernicus gave was misleading: it suggests a much greater simplicity than he indeed could effect. It shows the sun in the middle of the planetary orbits, but when he abandons general considerations and enters into astronomical details, Copernicus has to assume the centre of the earth's orbit (which does not coincide with the sun) as the centre of their movements. Moreover, the picture as he gives it may look quite simple, but in fact it dispenses with only those circles which stand for the orbit of the earth. As long as the earth had been supposed to stand still, the mathematical hypotheses for all other heavenly bodies had to take into account, not only their own motion, but also that of the earth. Moreover the planets, the sun and the moon in fact have elliptical (not circular) orbits. In order to account for the phenomena ensuing from this situation, Copernicus had to introduce a great many epicycles and even epicycles on epicycles. Finally there were in his system very slow movements (e.g. precession) of the earth's axis which require even more circles. The result is that, apart from the removal of the equants, there is no great simplification in the system as compared with the Ptolemaic one. According to Professor Otto Neugebauer 'the popular belief that Copernicus' heliocentric system constitutes a significant simplification of the Ptolemaic system is obviously wrong. The choice of the reference system has no effect whatever on the structure of the model and the Copernican models themselves require about twice as many circles at the Ptolemaic models. . . .'

Nevertheless, Copernicus believed that he had discovered 'the motion of the machinery of this world which has been built for us by the best and most orderly Craftsman of all'.

If Copernicus' system was not 'a significant simplification of the Ptolemaic system', what was its basic difference? Why was it so complicated?

Copernicus put the sun into the middle of the universe and he removed the equant which the Ptolemaic system used. Because he remained wedded to uniform circular motion he had to use many epicycles to account for the irregular motions.

3.6 Theological opposition

Besides the philosophical or physical critique Copernicus expected another not less weighty attack on his theory, from those 'idle talkers' who, ignorant of mathematics, 'shamelessly distorting the sense of some passage in Holy Scripture', would condemn his work. He reminds his readers (and foremost Pope Paul III) of the stock example of such misguided people, Lactantius, 'otherwise a distinguished writer but hardly a mathematician', who 'speaks in an utterly childish fashion concerning the shape of the earth', laughing at those who affirmed that the earth has the form of a globe. We need not be surprised, he says with disdain, if people like that laugh at us, but 'mathematics is written for mathematicians'.

If Copernicus had illusions that his fellow astronomers would be convinced by his demonstrations, he seems to have had less hopeful expectations about theologians and the multitude. At any rate, whereas friends like Nicholas of Schönberg and Bishop Tiedemann Giese had been rather optimistic, Copernicus expected at best an open mind in the Pope to whom he dedicated his book. Rheticus deemed it necessary to write a work (now unfortunately lost) to demonstrate that the theory of the motion of the earth is not in conflict with the authority of Scripture, whereas Osiander, who was the final editor of *De Revolutionibus*, added, against Copernicus' wish, a Preface intended to placate philosophers as well as theologians, who would feel offended by a doctrine so manifestly 'absurd', 'unphilosophical' and 'anti-scriptural'.

Osiander's Preface could hardly deceive the readers of Copernicus' work, but it would certainly remind them of the tenet that astronomical or mathematical hypotheses need not be true, or even probable, and that one could skip Copernicus' proofs for the physical truth of his theory. But they could admire his use of these 'absurd' hypotheses to fulfil so well the task of the astronomer: to 'save the phenomena' by devising hypotheses which would reduce all apparent movements to divine uniformity and circularity, and which would enable him to compute celestial movements in agreement with observations.

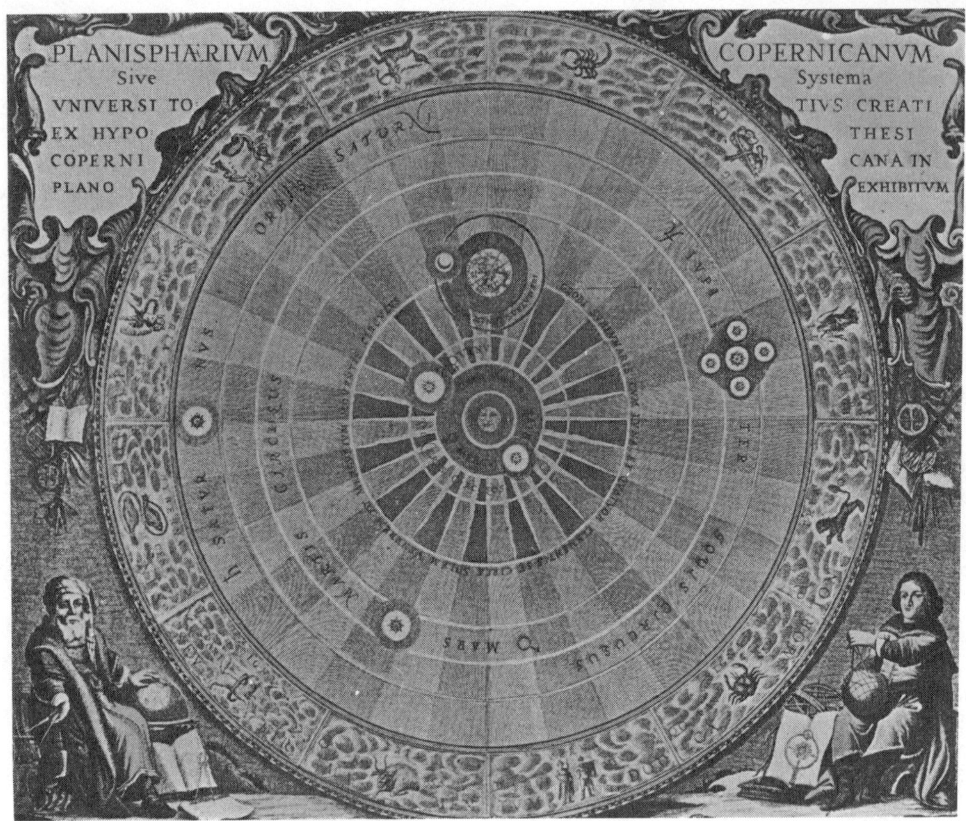

Figure 6 *The World System according to Copernicus (to which the Jupiter satellites discovered in 1610 have been added) from an original engraving (Mansell Collection).*

3.7 The world view of Copernicanism

Many writers on Copernicanism tell us that one of the greatest theological obstacles to the new system was that there is a conflict not only with particular Bible texts but also with the whole Christian world view. The Bible, so they say, takes an anthropocentric viewpoint. All other creatures are considered to exist for the sake of man and are submitted to his rule. It is for man's sake that Christ came into this world. In such a man-centred conception it goes without saying that man's dwelling is in the centre of the universe and that all heavenly bodies should turn around the earth. But what does Copernicus do? He reduces the earth to a mere point in the universe and he makes her a planet amongst the planets swerving through an immense space. Man, who formerly felt himself the central person, the king of the universe, now becomes an insignificant inhabitant of a not so important planet. No longer can it be maintained that all things have been made for man. Small wonder, then, that it should be said that people felt lost and annoyed because their secure position had been taken away.

The following quotation from a recent work should be compared with the extract from Copernicus himself:

In particular, man was *demoted* from the center of the universe to a spinning, peripheral planet. Man's uniqueness and the idea of God's particular concern for him seemed in danger. . . . The new cosmology was resisted, then, not only because it challenged the authority of Aristotle and Scripture, but because it threatened the whole Aristotelian sphere of purpose and meaning in which man's spatial location was correlated with his status in the cosmic hierarchy.[1]

So it is also as to the place of the earth; although it is not at the centre of the world, nevertheless the distance [to that centre] is as nothing, in particular when compared to that to the fixed stars.[2]

1 **What kind of source is the first extract likely to be?**
2 **Compare the first quotation with that of Copernicus and note what is said about man and the universe. What is your conclusion?**

Barbour is in fact a tertiary source (at least in this passage) and the book fits the criteria discussed in Unit 1 p. 21. His quotation seems to have perpetuated a common misunderstanding, namely that the removal of man from his central position had profound metaphysical significance for Copernicus. In fact Copernicus says that, because of the immense size of the universe, man's displacement is a relatively trivial matter, and he still is virtually at the centre. Copernicus stressed that the earth, and even its orbit, is a mere point in the universe. But his medieval predecessors (like the extremely popular John of Holywood (or Sacrobosco)), also spoke of the earth as a mere point, and even Ptolemy had said the same. Moreover, in the Aristotelian world view the sublunar terrestrial world was considered as liable to change, in contradistinction to the unalterable heavens. And all medieval theologians would lament that nothing is stable in this vale of tears. Philosophers and theologians vied with each other in inculcating into man's mind his own insignificance in comparison with his Creator, and the insignificance of his dwelling place in comparison with the size of the whole of creation. Copernicus, by extending the sphere of the fixed stars, could hardly have increased man's sense of being lost in the vastness of the universe.

Moreover, it should be realized that there always were two evaluations of man's position, and in the period immediately after Copernicus theologians and philosophers continued to emphasize also the uniqueness of man in the universe. Though he is but dust (Genesis 3: 19) he is the only creature made in God's image (Genesis 1: 26–7). This paradoxical situation, most clearly expressed in the eighth Psalm[3], even the idea of an infinite universe could not alter.

Man, as Pascal said, is but a point in an infinite universe that swallows him up, yet he surpasses the universe because he knows that he exists and he comprehends the universe by thought. Copernicus was completely in tune with such a conception. On the one hand denying to the earth a central position, he attributed to her on the other hand a rank which put her on an equal footing with heavenly bodies. And he placed the sun in the middle of the universe so that this lamp might illuminate everything in the whole 'machinery of the world', 'which has been built for us by the best and most orderly Craftsman of all'.

Similarly Kepler, who was a staunch defender of the Copernican world system, said that 'all things have been made for man'. Another defender of Copernicus, Philip van Lansbergen, said that 'all things exist to the glory of God and for the benefit of mankind'. In his opinion, the location of the earth's orbit halfway within the sequence of planetary spheres meant that the great heat of the sun was tempered by the spheres of Mercury and Venus. Evidently anyone who believed all things are for the sake of man, whether he was an Aristotelian or a Copernican, could easily find arguments to back this up.

At any rate, if Copernicus had not an anthropocentric *world picture*, he had, no less than anti-Copernicans, an anthropocentric *world view*. Copernicus did not

[1] I. G. Barbour, *Issues in Science and Religion*, S.C.M., 1966, p. 33.

[2] Nicholas Copernicus, *De Revolutionibus*, Book I, chapter 6.

[3] Please look this up.

whisper a word suggesting opposition might be expected to his displacement of man from the centre of the universe and reducing his home to a small speck. Copernicans did not have a religious world view different from that of anti-Copernicans.

In a satirical essay of John Donne (1571–1631) against the Jesuits, 'Ignatius his Conclave' (1611), Lucifer looks for a lieutenant who should have introduced novelties in religion and thereby proved himself an enemy of the ancient Biblical doctrine. Several candidates present themselves, amongst them Paracelsus, who had substituted the Aristotelian four elements with his three chemical principles, and also Copernicus. Copernicus pleads that he has raised the earth, whose centre is Lucifer's prison, into the heaven, and put the sun, the enemy of the Prince of Darkness, into the lowest part of the world. Then Ignatius of Loyola interferes and asks:

> But for you, what new thing have you invented by which our Lucifer gets anything? What cares he whether the earth travels or stands still? . . .
>
> Do not men believe, do they not live just as they did before? Besides, this detracts from the dignity of your learning and derogates from your right and title of coming to this place, that those opinions of yours may very well be true. If therefore any man have honour or title to this place in this matter, it belongs wholly to our Clavius, who opposed himself opportunely against you and the truth, which at that time was creeping into every man's mind.

Now it is a risky thing to write history in a deductive way, the more so as history, just like science, has to keep to data and facts. The historian has to be aware of his own prejudices. To highlight some of the problems extracts are given below from a recent enquiry published in the monthly periodical *Poland* (1972, no. 1). Read these and then answer the following questions.

A   The Copernican revolution *based on the natural sciences* transformed the picture of the world as it existed in Antiquity and the Middle Ages into a modern one. The reception of the Copernican system and the spreading of modern scientific thought stemming from the Copernican revolution, proceeded over many centuries and constituted part of the struggle between the materialistic and idealistic outlook. The sharp edge of this struggle was above all directed against *theology* and the *late scholastic philosophy* and educational theories, and finally also against the theistic positions of idealistic philosophers of a later period.[1]

B   The hypothesis to place the Sun in the centre of the Universe not only seemed to contradict observation but also opposed the accepted ideas and the image of the world *made sacred by the Bible*, but above all the concept of the *special position* occupied by the Earth.[2]

C   By depriving Man of his central, privileged place Copernicus disturbed the *traditional humanism*, but simultaneously renewed it on the cosmic level, giving first place to the *unity and solidarity of mankind* in the face of the Universe, which has to be studied and conquered.[3]

In the light of this section comment on the phrases in italics.

based on the natural sciences: The 'result' is here confused with the 'cause'. The Copernican system which took so long to be accepted was one of the major factors in the movement which led to the emergence of natural science. Indeed Treder seems tacitly to acknowledge this in lines 3–4.

theology: The Copernican System contradicted the Aristotelian world picture but can scarcely be said to be directed against theology. Copernicus was, after all, a Churchman and took pains to have his book accepted by the Pope.

[1] Prof. H. J. Treder, East Berlin, p. 40.

[2] Prof. A. Kauffeldt, Magdeburg, p. 46.

[3] Prof. J. Fabre, Paris, p. 44.

This whole passage A bears a remarkable resemblance to the writings of Draper and White discussed in Unit 1. It stands in very much the same tradition and shares the same weakness.

late scholastic philosophy: (see Introduction, also *Hooykaas* (SET BOOK), p. 7 and pp. 31–2). There is truth in this; Copernicus stands in direct contradiction to the Aristotelian and neo-Platonic cosmological views of the Middle Ages, whereas his physical conceptions, though showing the influence of the older philosophies, deviate substantially from them.

made sacred by the Bible: The image of the world against which Copernicus' system was set was essentially Aristotelian and was Biblical only in the sense that the Bible was read through Aristotelian spectacles.

special position: Copernicus believed he was enhancing rather than diminishing this 'special position' of the earth.

traditional humanism: Copernicus felt he was exalting the earth and man rather than debasing them.

unity and solidarity of mankind: This sounds too modern to be true!

We have to explain the fact of theological opposition on a plausible supposition which should be corroborated by some data at least. There is also the fact that a growing number of theologians accepted Copernicanism, and it would be quite easy to explain this on the same lines as the opposite opinion. If the Word became flesh not in Imperial Rome but in the insignificant land of Palestine, and then not in the capital, Jerusalem, but in the small town of Bethlehem, and if He lived not as a Prince or a High Priest in the capital but as a carpenter in despised Nazareth, it would be an edifying thought that He was *not* born in the centre of the universe but on the small planet earth, just an outpost of the world. Nobody (as far as we know) has put forward this 'plausible explanation'; it is not corroborated by facts, and consequently, however plausible it may seem, it has to be discarded. But if there had been a general approval of the Copernican theory in theological circles this explanation could have been offered, and as easily accepted as the one put forward so often to the contrary. All of which is simply to remind ourselves to be wary of facile generalizations in history.

3.8   Biblical exegesis and the motion of the earth

3.8.1   The problem

In his Preface to *De Revolutionibus*, Osiander's fear of theological opposition arose not because the Copernican system went against the Biblical anthropocentric world view, but because of the way certain specific Biblical texts were interpreted.

Until the eighteenth century it was a universally accepted belief that God reveals Himself to mankind in two 'books', Nature and Scripture. Scripture was considered the clearer revelation of the two and it was recognized by all Christians as a collection of divinely inspired writings. Its authority was fully accepted by the Roman Catholics, Greek Orthodox and Protestants alike.

However, once this basic belief was accepted, difficulties arose. Some words in Scripture seemed to be intended literally (for instance that Christ had risen on Easter morning), others less obviously so, for instance that God is angry, that He repented for what He had done, or certain passages that speak of God's hands, eyes and ears.

To many people today such passages appear to be largely metaphorical, or symbolic, in character. But, again, we have to dissociate our ideas from those of the past. Can we read this attitude in (say) writings of the seventeenth century? Extract 2 in *Anthology I*[1] is by the astronomer Kepler, in the introduction to his *Astronomia Nova*.

---

[1] D. C. Goodman (ed.), *Science and Religious Belief: A Selection of Primary Sources*, John Wright and Sons/The Open University Press, 1973 (ANTHOLOGY I).

Kepler does not speak of symbol or metaphor but rather regarded Biblical language about nature as making *accommodation* to popular belief and not necessarily being 'true-in-itself'. For him, at least, interpretation of such Biblical passages required judgement as to their ultimate purpose, *viz.* to convey religious rather than scientific truth.

In earlier centuries the Church Fathers had, with some difficulty, adapted their interpretation to fit in with the Aristotelian world picture. Consider, for example, the Genesis reference (1: 7) to 'waters above the firmament' (or heaven). These could not be clouds since the text indicated they were above the heaven. Origen identified them as 'angels' but this was too far-fetched for most.

Augustine (354–420) insisted that here and elsewhere the text will have not only a literal but also an allegorical meaning (e.g. that the 5 loaves in the story of the feeding of the 5,000 signified the 5 books of Moses). Moreover, apart from the question whether certain texts should be considered as being literally true, or as having but a figurative sense, the general problem arose whether the authority of Scripture *in itself* or of Scripture as understood by Tradition, was limited to matters of doctrine and ethics, or whether it also extended to matters of, for instance, cosmology.

Most people believed that the general outline of the Aristotelian world picture was to be found in Scripture. Consequently the hypothesis of the daily rotation of the earth was regarded as being in conflict with Biblical texts (as with Aristotelian theory) even by the fourteenth-century French philosopher and bishop Nicole Oresme, perhaps of all medieval scholars the one to go furthest in his positive appreciation of the hypothesis of the daily motion of the earth. He recognized that, mathematically speaking, the daily motion of the earth fitted the phenomena equally as well as the opposite current opinion. Against arguments borrowed from Aristotle's physics that deny this motion, he put counter-arguments *for* it, largely borrowing from the Platonic conceptions which Copernicus, too, was to use 150 years later. He even pointed out that many texts from Holy Scripture appearing to oppose the idea of the motion of the earth may be conceived as an accommodation to common speech.

In 1546, the Council of Trent decreed that, when interpreting the Bible, one should not deviate from the consensus of the Church Fathers. Moreover, as scholastic philosophy was re-established, a serious limitation to liberty of interpretation appeared. This was a weapon to be used against the opinions of the Reformers, as well as against divergent interpretation within the Roman Church. Two views then have to be taken into account: (1) the Bible is a source of scientific data, and any Biblical statement with cosmological implications should be taken as scientifically true; (2) the Bible is not meant for scientific information but rather accommodates itself to naïve observation, and sometimes even to wrong ideas prevalent in the time of the authors.

Anti-Copernicans, who had a long tradition behind them which had in some way harmonized Biblical interpretation with the current Aristotelian world picture, showed a tendency to take the first standpoint. The Copernicans on the other hand had no consensus of Fathers behind them and were more inclined towards the second point of view. Nevertheless, nor could they sometimes resist the temptation to derive Copernicanism from Biblical texts, as for instance the German author J. J. Zimmermann who wrote a book *Holy Scripture Copernicizing . . . or Astronomical Demonstration of the Copernican Structure of the World from Holy Scripture*, 1690.[1] A parallel case is that of those followers of Paracelsus who tried to back up their doctrine of three or five elements by reference to the book of Genesis.

Before answering the following read *Hooykaas* (SET BOOK), pp. 114–17.

---

[1] And even Galileo (*Hooykaas*, p. 124).

In what ways did the Biblical world picture seem to be 'distorted' by the anti-Copernicans?

Perhaps the chief ways were as follows:

1 A mistaken belief that their Aristotelian world picture was contained in the Bible.
2 Literalistic view of Biblical texts which can be taken otherwise as 'accommodation to common speech'.

3.8.2 The principle of accommodation

The principle of accommodation (which asserts that Biblical writers 'accommodated' their language to everyday usage) was taken much further by the Swiss Reformer John Calvin. See *Hooykaas* (SET BOOK), pp. 117–24.

On the basis of this passage:

1 **What was Calvin's attitude to Greek cosmological ideas?**
2 **How did Calvin regard those passages of the Bible which appeared to contradict these cosmological views?**
3 **How did those influenced by Calvin explain differences between natural/astronomical discoveries and Biblical language?**

Calvin accepted the Aristotelian world picture, but where it did not agree with the words of the Bible, he took the Biblical form as being that of common usage rather than 'scientific' description. 'Moses wrote in a popular way. . . . One should not look there for astronomy and other abstruse sciences; it is a book for laymen'.

He stressed the Reformation principle that, in Scripture, God speaks directly not only to the learned but also to the unlearned. Consequently the Bible gives no scientific information but accommodates itself to the way of speech of all, and the naïve conceptions of everyday life. It is remarkable that he did not reject the doctrine of the motion of the earth on theological grounds. It is not mentioned in his commentaries on texts with cosmological implications but in a sermon in which he considered the tenet of the motion of the earth as an offence against common sense, no less perverse than that of the ancient philosopher who said that snow is black. It goes without saying that if the Ptolemaic system is acceptable in spite of being in conflict with the literal interpretation of the Bible text, one cannot reject the Copernican system on the grounds that it is *not* conformable to the Biblical text. Small wonder, then, that Calvin's exegetical principles were applied by some Copernicans who were influenced by his teaching, e.g. Edward Wright (in his Preface to Gilbert's book on the magnet in 1600), and Philip van Lansbergen (1619 and 1629), a reformed minister in Zeeland who was one of the most ardent defenders of the Copernican system at that time. Kepler's argument on this topic completely agrees with that of Calvin. Dr Jacob van Lansbergen, in defending his father's books, explicitly mentioned 'our Calvin' as the source of his exegetical arguments.

John Wilkins, a great protagonist of Copernicanism and of scientific freedom in general, also appealed to Calvin's exegetical principles in order to ward off the criticism levelled against Copernicus' theory by adherents of literalist interpretation of the relevant texts.

In his *Discourse concerning a new planet tending to prove that it is probable our earth is one of the planets* (1640), Wilkins tried to 'remove those common prejudices which usually deter men from taking any argument, tending this way into their consideration'. He objected to the arguments which the priest Fromondus had adduced against Philip and Jacob van Lansbergen. It is evident that he borrowed many of his counter-arguments from the Zeelandish authors (though he did not say so) but he penetrated more deeply than they had done into the problem of Scripture and science. He argued that the places in Scripture which seem to intimate the diurnal

Figure 7 Dr John Wilkins (Courtesy of the Warden and Fellows of Wadham College, Oxford, Photo by B. J. Harris).

motion of the sun are capable of another interpretation; that the Holy Ghost in many places of Scripture plainly does not speak of things as they are in themselves, but as they appear to us; that diverse learned men have fallen into great absurdities while looking for grounds of science from the Scriptures; that Scripture, rightly understood, does not affirm the immobility of the earth or its place in the centre of the universe.

Wilkins held the opinion that practically all astronomers of note and skill were Copernicans. The anti-Copernicans vigorously denied this and it may be that Wilkins' verdict was determined by the exceptionally favourable situation in England. He said that Scripture is infallible, even when speaking on natural things, but that its words should be taken in the sense in which they were intended, that is as accommodated to vulgar opinion. Calvin's commentaries were extensively quoted, but he has also examples not borrowed from Calvin or Lansbergen.

He pointed out that when Scripture speaks of the stars that shall fall from heaven (Mark 13: 25; Revelation 6: 13), this was to be interpreted, not of true stars, but of those that appear so, namely meteors. Scripture speaks of common natural effects as if their true causes were altogether inscrutable because they were generally so esteemed by the vulgar, as for instance when it is said that none know whence the wind comes or whither it goes (John 3: 8).

Wilkins then attacked the ancient opinion of Jewish and Christian writers who held that Scripture not only contains things which concern our religion and morals but every secret in any art or science. Those who examine Biblical texts by the exact rules of science commit the same fault as St Basil, who held that the moon is bigger than any of the stars because Moses calls it a great light; they commit the same fault as most Church Fathers who maintained that there are waters above the starry firmament, and as St Chrysostom who doubted that the heavens have a spherical form because Scripture compares them with a tent (Isaiah 40: 20 and Hebrews 8: 2).

Wilkins was quite impartial, for he said that the opinion of many Church Fathers that only a miracle prevented the seas from overflowing the land (Jeremiah 5: 22) was shared by Thomas Aquinas, Luther and Calvin.

When he enumerated the diverse arguments from the words of Scripture, the principles of nature and the observations in astronomy generally put forward to prove that the earth is in the centre of the universe, he said nothing about the 'religious' argument invented by our modern writers concerning the lost preeminence of this dwelling-place of mankind. But he does mention the argument of Fromondus who holds that, as the heaven is the seat of the blessed, the earth must be most remotely situated from it, namely in the centre of the world. Of course Wilkins would not accept this argument as having any validity.

Which of the two arguments that Copernicus expected to be used against his theory, were in fact used by Calvin in his opposition to the motion of the earth?

Calvin did not mention its incompatibility with Scripture when discarding the idea of the motion of the earth; he rejected it because he held that it was contrary to common sense; it was 'absurd'.

3.8.3 Literalistic interpretation

However, not all sixteenth-century astronomers accepted a 'reconciliation' between Copernicus and the Bible. Foremost amongst those who did not was the Danish astronomer Tycho Brahe (1546–1601). To his credit are the most exact observations in astronomy before the introduction of the telescope. He had some astronomical reasons for rejecting the motion of the earth, but a decisive influence was his literalistic interpretation of some Bible texts (1583). In his own astronomical system the planets revolve around the sun and the sun with all its satellites revolves around the earth which stands immovably in the centre of the universe. Even the daily rotation belongs to the heavenly bodies and not to the earth.

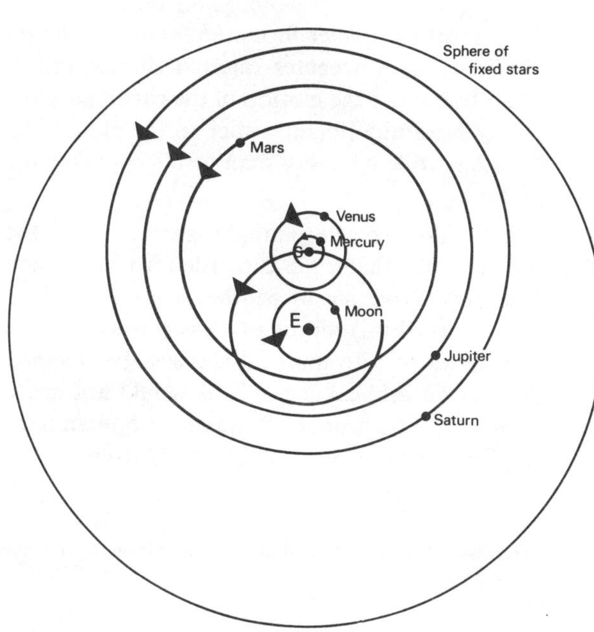

Figure 8   Portrait of Tycho Brahe (Mansell Collection).

It was certainly not Aristotelian physics that prevented Tycho's acceptance of the motion of the earth. He had little respect for authorities in science and undermined Aristotle's physics of the heavens by demonstrating that the new star of 1572 did not belong to the sub-lunar regions but was farther away than the farthest planet, amongst the fixed stars. As this star arose and practically disappeared within a year and a half, it was evident that generation and corruption also reigned in the heavens, a notion totally opposed to the fundamental tenets of Aristotelian physics. Tycho's system became the refuge for those astronomers who did not want, or did not dare, to accept Copernicanism, and who also felt that the Aristotelian Ptolemaic system was unsatisfactory. It was to play an important role in the seventeenth century as an alternative to Copernicanism.

The Tychonic system was often modified by taking one step further in the Copernican direction and admitting the rotation of the earth. This was done, for instance by

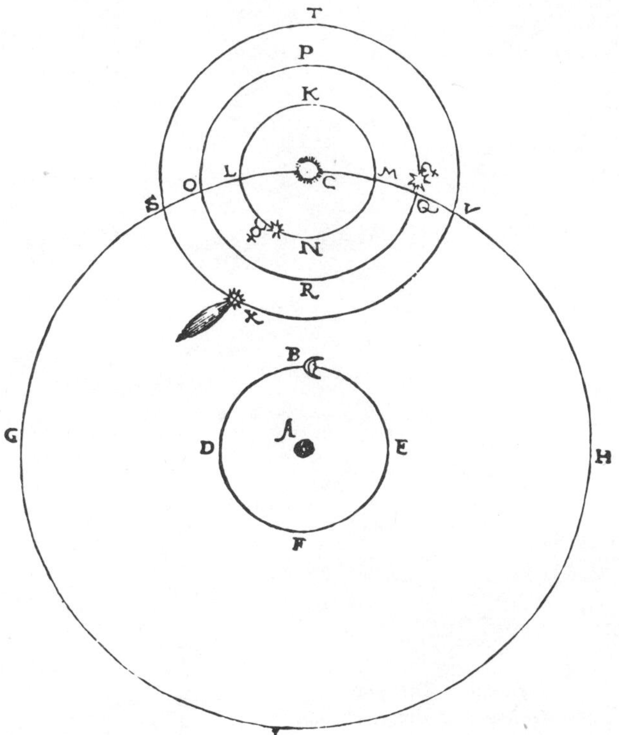

Figure 9   Tycho Brahe's planetary system showing at X the comet of 1577, from Tycho Brahe, De Mundo aetherei recentioribus phaenomenis *(Ronan Picture Library).*

Nathanael Carpenter in his *Philosophia Libera* (1622) and his *Geographie Delineated Forth* (1625). Carpenter rejected the arguments of those who refer to Scripture in order to oppose the motion of the earth. He calls them 'precise men who will condemn without examination and stick to the plain letter notwithstanding all absurdities'. But he wants to leave every man to his own free judgement to embrace or reject what he pleases.

It should be borne in mind that rejection of the Copernican system as a physical truth did not mean that it was discarded from astronomical teaching or computing. Nicholas Mulerius, Professor of Mathematics at the University of Groningen, was the editor of the last (third) edition of Copernicus' work in 1617 which was intended as the first in a series of astronomical classics. He deemed the book indispensable in teaching, and yet on Biblical grounds he would not accept the annual revolution of the earth. Similarly, Willebrord Snel was a Tychonian but nevertheless made models for teaching the Copernican system in Leiden University.

What were the common and contrasting elements in Tycho Brahe's system and Copernicus' system?

The planets revolve around the sun in both—but Tycho did not accept the movement of the earth or that it was one of the planets. The earth was the centre of the universe with Tycho, but the sun with Copernicus. He also denied that certain Biblical texts were merely common parlance, but interpreted them literally.

The Biblical issue became an important topic only after the Galileo affair. In 1616 the Roman Congregation of the Index stated that the Pythagorean doctrine taught by Copernicus and by Didacus a Stunica (Diego de Zuniga) in his commentary on Job (1584) had been spread by the Carmelite father Paolo Antonio Foscarini in a letter on 'the opinion of the Pythagoreans and of Copernicus concerning the motion of the Earth' (Naples 1615) who attempted to show that the sun's immobility and the earth's motion are consonant with truth and not opposed to Holy Scripture. It was decreed that the two first mentioned works should be corrected, whereas the latter should be altogether prohibited. At the same time Galileo got a warning to refrain from teaching the Copernican doctrine.

Figure 10   Title page of Galileo's Dialogue (*Ronan Picture Library*).

Foscarini had openly declared that he wanted to support Galileo and Kepler by proving that the motion of the earth is not against Scripture. Like Galileo he was of the opinion that the words of Scripture are adapted to the way of speech in daily life, and like Galileo he expected that a more penetrating exegesis of Holy Scripture would demonstrate that its authors had something like the Copernican theory in their minds and that Scripture would provide arguments for it.

After the publication of his *Dialogue Concerning the Two Principal Systems of the World* (1632) and his condemnation by the Papal Inquisition in 1633, Galileo and his friends were convinced that machinations of the Jesuits were the main cause of the calamities he met with. To his Protestant friend Elia Diodati in Paris, he wrote in 1634 that the Jesuit father Grienberger who sympathized with him had said that 'if Galileo had only known how to retain the favour of the Fathers of that [Jesuit] College [in Rome] . . . none of his misfortunes would have befallen him, and he could have written what pleased him . . . even about the motion of the earth'. Galileo then adds that his present and former difficulties did not arise from a certain opinion but from his being in disgrace with the Jesuits.

Though personal hatreds and jealousies may have played an important role in starting the process against Galileo, the verdict would not have been possible if there had not been also the sincere conviction of many of his opponents that he had undermined the foundations of the Roman Catholic Church. The Council of Trent had enjoined on interpreters that in mattters of faith and morals, Scripture should be interpreted in conformity with the Holy Church which alone can judge about its true meaning, and not to go against the unanimous consensus of the Fathers of the Church. Galileo now had not only to discover that he as a layman had to leave alone Biblical interpretation but also that the Inquisition interpreted the verdict of Trent as an injunction to keep to the opinion of the Church Fathers in *all* things. He was declared 'suspected of heresy, namely of having believed and held the doctrine—which is false and contrary to the sacred and divine Scriptures—that the sun is the centre of the world and does not move from east to west and that the earth moves and is not the centre of the world'. Galileo had to say 'with sincere heart and unfeigned faith I abjure, curse and detest the aforesaid errors and heresies'. How little such an oath means becomes evident from his correspondence with his friends in Italy, France and Holland. Not a month afterwards he connived at a translation of his work into Latin by the Protestant astronomer Matthias Bernegger in Strasburg and its publication (in 1637) together with Foscarini's work. In spite of the prohibition to publish old or new works, the manuscript of Galileo's *Two New Sciences* was smuggled out of Italy and printed in heretical Holland (1638). Catholics in Italy and France blamed the Inquisition and the Jesuits ('the two scourges of the truth' as Pascal was to call them) for what had been done. Italian scientists henceforth worked on harmless topics, mathematics and experiments. 'I have sat among their learned men', wrote Milton (1644), 'and been counted happy to be born in such a place of philosophic freedom as they supposed England was, while they did nothing but bemoan the servile condition into which learning amongst them was brought; that this [namely the Galileo affair] was it which had damped the glory of Italian wits'.

Many Catholic scholars, and among them several outstanding Jesuit astronomers who conformed to the decision of the Holy Office, now defended not the Ptolemaic but the Tychonian system, as did also the conservative Protestants especially in Germany and the Scandinavian countries.

3.9   The later reception of Copernicanism

The Biblical issue became an important topic only after the crisis associated with Galileo. This is so important that it forms the subject of the next unit. Meanwhile, however, it would be worthwhile to read *Hooykaas* (SET BOOK), pp. 124–6.

The revival of scholastic philosophy by the Jesuits, with some non-essential changes to bring it more up to date, had an enormous influence on the teaching not only in the Roman Catholic but also in Protestant universities on the continent. It should be realized however that the younger reformers, Melanchthon in Wittenberg and Beza in Geneva, had already reverted to Aristotelian philosophy. In Holland the theologian Voetius (1636) kept to scholastic philosophy as well as to a literalistic interpretation of the Bible. To beginners in theology he recommended the commentary by the Portuguese Jesuit Father, Benedict Pereira. He was aware of the fact that Calvin's commentary would contradict his own way of interpretation. In his opinion the Bible was a source for all sciences, astronomy included. He deemed it impossible that scientists would ever discover whence the wind comes and whither it goes, because the Bible text states that man did not know it (which was certainly true in Voetius' time). If the Holy Spirit would have said that the earth stands still only in order to accommodate Himself to the common people and not because it is objectively true, this would mean that He would have told a lie on behalf of the common people.

Voetius had a wide influence in theological circles but he had no legal authority to impose his ideas on anybody and he had no theological monopoly in his church. People in Holland and England could choose their philosophical standpoint in freedom. If the official prescription was that scholastic philosophy should be taught in the university, this did not mean that the Professor was obliged to keep to it in his writings and conversations. Voetius' own university of Utrecht was the first where Cartesianism was started and where a strong anti-scholastic group of theologians and philosophers favoured Copernicanism and Cartesianism. All parties, Aristotelians, Tychonians and Copernicans were represented in the republic of the seven United Provinces.

In the same way in England, all sects of philosophy, the Copernicans included, were represented in the universities. As Wilkins declared, the leading scientists were 'ready to follow the banner of truth by whomsoever it shall be lifted up'. And Seth Ward said that practically everybody who knew something about astronomy accepted the Copernican system either as true or as the most convenient hypothesis (1654).

The resistance to Copernicanism in England had always been very weak. Already in 1576 the mathematician Thomas Digges, who was of a Puritan family, put forward his *Perfit Description of the Caelestiall Orbes* which contained a free translation of the first book of Copernicus' *De Revolutionibus*. It was illustrated by the first picture of a universe without definite bounds. Opposition to Copernicus on Scriptural grounds did also occur (William Barlowe 1616, Thomas Tymme 1612, Alexander Ross 1634, John Owen 1671), but the protagonists of Copernicus were more numerous. This has certainly some connection with the fact that in no other country was the 'new philosophy', that is corpuscular mechanistic science, so widely accepted. At the other end of the scale were Spain and Portugal where the Inquisition was more strict than in the Papal State and where Copernicanism remained non-existent, while in England and Holland it was already generally accepted.

In France outstanding scholars like Roberval and Pascal, and the priest Boulliaud, more or less openly declared Copernicus' doctrine, if not strictly proven, at least probably true. Boulliaud in 1639 anonymously published a pro-Copernican treatise in Amsterdam, but in 1645 he published another one under his own name in Paris.

One of the strangest cases is that of Descartes. He wrote his main works during his long stay in Holland. Yet he was greatly embarrassed by the condemnation of Galileo. He was of the opinion that, if Copernicus were wrong, the foundations of his own philosophy were also wrong and he withheld the publication of his *Le Monde* (1633) for fear of giving offence to the authorities of his church. In 1644, however, in his *Principles of Philosophy*, he triumphantly announces 'that I deny the movement of the earth, more carefully than Copernicus and more truthfully than Tycho'. Yet he then unblushingly advances the Copernican system. Motion, he says, is always a relation to the immediate surroundings. When the earth turns round, the lowest layer of the surrounding ether vortex has the same velocity as the earth's surface. Consequently the earth then does not move. Moreover the earth's vortex is swept along in the larger ether vortex of the sun. So in its annual revolution it does not move either, i.e. not with relation to the solar vortex. This reasoning boils down to the statement: the Holy Office does not allow me to say that the earth turns round, but it does not say anything about an ether vortex turning round.

In the seventeenth century, as we shall see in Units 4–5 and television programme 4, mechanistic philosophy gradually supplanted Aristotelian scholasticism and the

Figure 11   René Descartes, 1596–1650, plate from Opera Philosophica (1656). The poem is by Constantijn Huygens after an etching by Francis van Schooten, a mathematician of the Cartesian circle (from a print in the Royal College of Physicians).

Platonist–Pythagorean Renaissance philosophies. The character of a substance was believed to be determined by the size, shape, arrangement and motion of its smallest particles. All changes should be thought ultimately in terms of movements, impacts and rearrangements of these particles. The system admitted neither teleological arguments (change has not to be explained by its purpose but by its mechanical cause), nor aesthetic arguments (like beauty, order and symmetry of arrangement).

Though seventeenth-century mechanistic philosophy borrowed its concepts from ancient Greek corpuscularian theories of nature, it should not be considered as a mere revival of the atomistic philosophy of Antiquity. Ancient materialism assumed blind Fate would rule all events. *Mechane* or *Machina*, however, means a clever device: in the mechanistic philosophy the world is like a clock made according to divine plan.

The mechanistic philosophers had no difficulty at all in inserting the Copernican system into their world picture. They needed neither the physical arguments of the Copernicans for it, nor those of the anti-Copernicans against it. In general they had no predilection for the conception of a closed universe and they considered the sun as one of the fixed stars. Neither the earth nor the sun was considered to occupy the centre of the universe. Most mechanistic philosophers rejected the idea that one thing or one place is superior to the other.

The introduction of the telescope helped to undermine the idea of hierarchical order, at least in its scientific aspects (see television programme 2).

In conclusion we may note that seventeenth-century mechanistic philosophy, though in itself neutral, would also tend to the Copernican system especially after Newton's synthesis. It was, however, not always a sign of intellectual backwardness that a scholar adhered to the old system; neither for mathematical reasons (computation founded on observations) nor for physical reasons (conformity to a general view of nature) could a strictly cogent decision be made before 1700. Consequently, many Copernicans claimed high probability, not absolute certainty, for their system. Many others, however, known to have been in sympathy with the Copernican system, submitted to what they considered freely, or on the authority of their spiritual leaders, as the explicit teaching of Scripture, which prevented them from accepting the motion of the earth. But to suppose that there should have been a spiritual upheaval because man was dethroned from his allegedly central position in the universe is to follow a myth.

4 The Plurality of Worlds

4.1 The analogical argument

The great problem that now loomed large was an immediate consequence of the fact that analogies may be drawn in opposite directions. If some known properties of A closely resemble some properties of B, it may be that some other properties of A are analogous to properties of B as yet unknown, or that some already known properties of B may correspond to as yet unknown properties of A. Thus Jupiter is a planet and it has moons (Galileo); the earth has a moon; the earth may be a planet too. But also conversely: the earth is a planet and is inhabited; Jupiter is a planet; Jupiter may be inhabited too. In the same way: if the fixed stars are suns, we might expect them to have planetary systems or worlds with them. Consequently Copernicanism inevitably gave rise to two new problems: first, that of the inhabitability of the planets and of the moon (which is a planet of a planet), and secondly, that of the plurality of worlds.

From what you have read so far can you suggest what theological issue underlay these two problems?

Theologically speaking, both problems were closely related because both of them raised the question as to whether there could be other rational beings besides Adam's offspring. It was just as obvious then as at any time since that Scripture does not give the slightest indication on the matter. The story of creation in the Book of Genesis only speaks of man made in God's image and placed on this earth. But if there are other inhabited worlds, instead of saying that God made all things for man it could have been possible to hold that He made the earth for man, and perhaps other inhabited bodies for the sake of their own inhabitants. It is true that this issue lay outside the scope of human science. It seems, however, that man has an irresistible urge to put questions that he is not, at least not yet, able to answer and that he often anticipates those answers 'in the name of science'. Against this anticipation there is no objection, scientifically speaking, provided the answer shows an awareness of the limited probability of its truth. As usual, the theological problems were here connected with philosophical ones. Not only does Scripture fail to mention other inhabitable planets or the plurality of worlds, but Aristotelian philosophy was even explicitly against it.

4.2 Plurality before Copernicus

Aristotle had tried to show that only one world is possible, that is one central body surrounded by planetary spheres. Materialistic philosophers like Democritus, Epicurus and Lucretius held that there are many worlds. In the ancient controversy, plurality of worlds meant that *the whole universe* is repeated many times and not, as in later theories, that in remote corners of this our universe there exist suns and planetary systems which may sometimes contain inhabited planets.

In the Middle Ages the idea of the plurality of worlds, in the sense of a multitude of Aristotelian spherical universes, was vigorously denied by Thomas Aquinas (for instance) not only on theological but also on philosophical grounds. If there be more worlds than one, they must either be of the same or of a different nature. If they are similar, they would be superfluous, which is not according to divine wisdom; if they are different, our world would not contain everything possible, it would not be perfect and only the totality of those worlds would constitute one perfect universe. The Cardinal Nicholas Cusanus (1401–64) assumed that on the sun, the moon and the stars are living beings whose nature remains unknown to us, but must widely differ from ours. Cusanus did not accept a closed but rather an infinite universe.

In Unit 1, pp. 39 and 44, extracts from Simpson and Dijksterhuis refer to a problem that was closely analogous to that of the plurality of worlds. What was this other difficulty facing medieval theologians?

An analogous problem which had already given rise to theological and philosophical difficulties was that of the antipodes. Some philosophers and geographers of Antiquity had asserted that the tropics are uninhabitable. Theologians like St Augustine then drew the conclusion that the southern hemisphere could not be inhabited, for Adam's offspring could not have wandered through those fiery parts; a separate creation there was unacceptable, for it would then be impossible to obey the command of preaching the Gospel to all creatures, if there were some who could not be reached. In the eighth century Virgil of Salzburg got into serious difficulties because Boniface accused him of teaching that there *are* antipodes, and that another Saviour was necessary for them. In the Copernican controversy this story was triumphantly used against those who rejected data revealed by experience because of a literalistic interpretation of Bible texts.

After the Portuguese navigators had reached the equator, it was often emphasized that unlearned sailors had demonstrated as being true what the acute reasoning of philosophers had demonstrated to be impossible. This was a violent shock to traditional philosophy and to a traditional Biblical exegesis which had perhaps unwittingly been based upon it.

To medieval man the doctrine of plurality of worlds smacked of ancient materialism, a tenet of Epicurus which seemed reprehensible even without judging it on its own merits. In the sixteenth century Giordano Bruno, in the wake of Cusanus, propounded the doctrines of the plurality of worlds, the infinity of the universe and the inhabitability of other heavenly bodies. He ended his life at the stake in 1600.

The following is a list of the eight charges facing Bruno at his trial. How far do those substantiate the traditional view that Bruno was a 'martyr' to science?

The first bears 'on the generation of things' and on the affirmation of 'two principles real and eternal for existence: soul of the world and original matter from which things come to be.'

The second concerns the doctrine of the infinite universe and the innumerable worlds, which opposed the idea of a creation in time. For Bruno confirms, in the course of these interrogations, that 'whoever denies an infinite effect, denies infinite power.'

The third proposition is relative to the human soul. Deriving from the soul of the world, the human soul would be no more than a transitory phenomenon: all its reality would reside in the soul of the world, infinite and eternal; it is the same for all bodies, which are reducible to a material equally infinite and eternal.

The fourth proposition results from the first three; inasmuch as substance is eternal, nothing, therefore, can be created or destroyed; all transform themselves; individual lives and deaths are mere appearances: 'There is no mutation in substance,' but only in the particular forms which substance assumes.

The fifth proposition bears 'on the movement of the earth.' Bruno said he had demonstrated 'the cause of the terrestrial movement and of the immobility of the firmament' by 'certain reasons which do not bear any prejudice to the authority of the divine Scripture.' It would be in vain to show him the verses from Ecclesiastes (1, 4): *Terra autem in aeternum stat: Sol oritur et Occidit*; he would object that these words could be interpreted differently, and, in any case, that the Scripture is expressed in the language understandable by the faithful but is not addressed to the learned scholars as such.

The sixth proposition holds that the stars are veritable 'messengers and interpreters of the divine voice . . . the sensible and visible angels.'

The seventh proposition holds that the earth has a soul, 'not only sensitive . . but also intellectual' and perhaps more. Does not Genesis (1, 24) say *Producat terra animam viventem*, objects Bruno, as he aggravates his own case.

The eighth proposition concerns the individual soul and its relation to the individual body. According to Saint Thomas' teaching, which the Church adopted, the intellectual soul is the form of the human body. Bruno objects: 'I do not understand, in my philosopher's role, that the soul is a form, but that it constitutes an actual spiritual reality present in the body . . . captive in some way, in a prison . . .' as a sailor would be in a ship.[1]

In the light of these charges it seems that the Copernican view of the solar system was implied in his theory but had nothing to do with his condemnation. His pantheistic heresies gave more than sufficient reason to the Inquisitors for his condemnation. So radical was his theological position that monks were regarded as 'asses' and many Catholic doctrines as 'asinine'. Recent research has shown how impossible it is to maintain the thesis that Bruno was a 'martyr to science'.

Figure 12   Engraving by Bernard Picart of Cyrano de Bergerac from Les Oeuvres, *volume 1, 1709 (British Museum).*

Figure 13   Engraving by Bernard Picart for Cyrano de Bergerac, Les Oeuvres, *volume 2, 1709 (British Museum).*

[1] From Antoinette M. Paterson, *The Infinite Worlds of Giordano Bruno*, American Lecture Philosophy Series, Charles C. Thomas, 1970.

Figure 14   Headpiece to Cyrano de Bergerac, Les Oeuvres, *volume 1. Engraving by Bernard Picart* (*British Museum*).

4.3   Plurality and the mechanistic philosophers

In the following pages we will not enter into the theme of the inhabitability of planets or plurality of worlds as it was dealt with by writers of fiction like Francis Godwin (*The Man in the Moon*, published 1638) or Cyrano de Bergerac (*Voyage to the Moon*, published 1656) but will consider more scientific authors like Gassendi, Wilkins and Huygens.

Figure 15   Pierre Gassendi (1592–1655), engraving by C. Visscher (from a print in the Royal College of Physicians).

The French priest Pierre Gassendi, a mechanistic philosopher tending to Copernicanism, held that the moon would be inhabitable, but, because of the great difference of climate, its inhabitants must widely differ from us. He was of the opinion that the unknown plants, animals and rational beings of Mercury must be the smallest and the most developed, those of Venus bigger and less perfect, but more perfect than corresponding terrestrial beings, and so on. The sun must have the most perfect inhabitants because this globe is nobler than all the others. We see here that some 'hierarchy of beings' unavoidably crept into the mechanistic conception as it recognized a plan in the universe. Instead of holding that the universe has been made for man, however, Gassendi thought that we should say that God made it for His own glory, as He made angels for His glory. The charge that such a conception is impious is reversed: Gassendi deems it arrogant impiety to believe that God could not place rational beings on those globes and that they cannot be superior to us. It is self-adulation to think that God made all things for us. Is it not enough that He has honoured us by His visible presence, though we are but dust and ashes? Should we then refuse to admit that He could make creatures without any use to us? This shows that Christian orthodoxy sometimes found the habitability of other planets a highly edifying doctrine. It was also pointed out that God being infinite and incomprehensible, His creation, too, must be infinite and greater than the small part known by and allotted to man.

In his *Dialogue* (1632) Galileo put the question whether there are plants, animals, rain and thunder on the moon and the planets as on earth. He does not believe so, still less that they have human inhabitants. But if nothing similar to the terrestrial beings does exist there, this does not imply that there are no beings there that are liable to change, though our mind cannot conceive them.

Before 1620, Kepler wrote his *Somnium*, in which is expounded what one would see when performing the fantastic voyage to the moon described therein. In his fantasy,

Kepler describes the climatic and other conditions on the moon, and the way its inhabitants live. The underlying idea is that this earth has a starlike nature and that consequently the moon and the planets have a terrestrial nature. Kepler, however, was horrified by Bruno's opinion that the fixed stars were so many suns and that the universe would have no boundaries; he considered this a denial of the divine harmony and order of the world.

John Wilkins, on this subject also one of the most intrepid thinkers, in 1638 anonymously published *The Discovery of a New World, or a Discourse tending to prove that it is probable there may be another habitable world in the moon. With a discourse concerning the possibility of a passage thither.* The work was many times reprinted and it earned him lasting fame as his name was given to one of the spots of the moon by M. F. van Langren, who about 1650 edited one of the first lunar maps. Wilkins belongs to the new epoch in science. In spite of his sound humanistic training he does not look to the past but into the future. He was a Puritan who after the Restoration conformed and worked for reconciliation and tried to stem the reactionary tide.

Wilkins' advocacy of the plurality of worlds starts with the demonstration that the suppositions implied in this opinion do not contradict 'The Principles of Reason or Faith'. In Wilkins' opinion 'it is not Aristotle but Truth that should be the rule of our opinions' and he wonders why some should be so superstitious as to stick closer to him than to Scripture. But 'some will say that the position of a plurality of worlds is also directly against Scripture which speaks but of one world.' Wilkins' answer is that 'the negative authority of Scripture is not prevalent in those things which are not the fundamentals of religion'. This is an important and moderate statement, as there was a tendency amongst Puritans to admit only what was positively asserted in Scripture, that is: to find at all costs a Biblical support for anything they held.

It goes without saying, then, that Wilkins would not revert to Scripture for confirmation of his own belief in the motion of the earth and the plurality of worlds. When someone would reply that if there had been another world we should have some notice of it in Scripture, his answer is that it would be as probable that Scripture should have informed us of the planets, which neither Moses, nor Job nor the Psalms mention. 'It is evidently besides the scope of Scripture to discover anything unto us concerning the secrets of science.' And he then quotes 'our countryman Master Wright' and also Calvin who maintained that Scripture only speaks about things which are obvious to the common people as well as to the learned. In Wilkins' opinion one should follow St Augustine who said that when words of Scripture seem to contradict common sense or experience, they should be understood in a qualified sense and not according to the letter. On the other hand, other Fathers of the Church have fallen into absurdities for want of this rule, for instance when it was held that the earth is floating on water (because Psalm 24 says 'He has founded the earth upon the seas').

Wilkins would be ready to consent that there is one universe but he would not agree that there is no plurality of worlds, when 'world' is used in the same sense as when we call the earth a 'world'.

When answering the question of the plurality of worlds in the affirmative, Wilkins has to maintain that the heavens do not consist of unchangeable matter: their matter must be liable to the same change and corruption as our earthly bodies. This, then, seemed to be an unavoidable consequence of the Copernican theory and in this respect it was dynamite under the foundations of Aristotelian physics. Wilkins quotes Cusanus, and even the unorthodox Bruno and Campanella, as holding that there is a particular world in every star. 'A countryman of ours', Nicholas Hill, is quoted as saying that 'it is probable the earth has a starry nature'.

The invention of the telescope had aroused speculations on the habitability of the moon, which seemed to have mountains as well as seas (the dark spots), so that there must be some resemblance to the earth. Wilkins pointed out that not only Aristarchus, Philolaus and Copernicus, but also Rheticus, van Lansbergen and Gilbert affirmed the earth to be one of the planets. 'Now if our earth were one of the planets, then why may not another of the planets be an earth?' He does not deem it probable that the inhabitants of the moon are men like us. Probably they bear only some likeness to our nature, or perhaps they are beyond our understanding, which is limited by our senses. 'I dare not myself affirm anything of these Selenities, because I know not any ground whereon to build any probable opinion'. Evidently Wilkins is prone to keep to empirical certainties and to put forward only those speculations that have some probability in themselves. He hopes, however, that our posterity will find out a conveyance to 'this other world' in the moon and, if there be inhabitants there, will have commerce with them.

Figure 16   Headpiece to Le Bovier De Fontenelle's Entretiens sur la pluralité des mondes *in Oeuvres diverses, 1728, Engraving by Bernard Picart, 1727 (British Museum).*

No work on the plurality of worlds achieved such fame as the *Discourses on the Plurality of Worlds* (1686) by Fontenelle, the secretary of the Paris Academy of Sciences. The work is based on Cartesian physics which the year after was to receive a fatal blow by Newton. The author carefully avoids calling the inhabitants of other heavenly bodies men. As it was written in the light and elegant style beloved by the philosophers of the Enlightenment, it was a matter of discussion in French salons of that epoch (see television programme 3).

The greatest scientist among writers on the plurality of worlds was Christiaan Huygens (1620–95), a mathematician, philosopher and physicist of renown. He belonged to a patrician family which was closely allied with the House of Orange; his father had been secretary to the Stadholder Prince Frederick Henry, and his brother had the same function with Frederick's grandson, King Stadholder William III. Huygens' posthumously published *Cosmotheoros* (1698) seems of a purely scientific character as long as the author imagines how one would see the heavenly bodies when standing on one of the other planets, but he soon launches into speculations on the character and civilization of the hypothetical inhabitants of those planets in a way which is not at all characteristic of this sober-minded mathematical and experimental physicist. The *Cosmotheoros* was translated into French, English, German and Russian. Peter the Great admired it so much that he ordered it to be translated and published, so that it became the first Western European scientific treatise printed in Russia (1717). The director of the printing office was deeply shocked by its contents which he deemed atheistical and blasphemous. The Czar being abroad in Holland, the director compromised with his conscience by printing only a few copies. In 1727, however, Peter ordered a second printing.

According to Huygens, a Copernican who holds that the earth is a planet 'cannot but sometimes have a fancy, that it is not improbable that the rest of the planets have their dress and furniture, nay and their inhabitants too as well as this earth of ours', especially when he considers the later discoveries of the satellites of Jupiter and Saturn. The telescopic observation showing that the Moon has hilly regions is an argument for the relation between the moon and the earth. Huygens reminds his readers of Cusanus, Bruno, Kepler and Fontenelle who furnished the planets or even the sun and the fixed stars with inhabitants. This, however, 'was the utmost of their boldness'. He himself wants now to go further and to enter into conjectures on those inhabitants. He warns, however, that he does 'not assert anything as positively true, for this would be madness'; he is only advancing a probable guess. Some people will claim that we set up conjectures against Holy Writ which does not say one word on the production or existence of such creatures. To them the answer is that God evidently had no design to enumerate all the works of His creation in the Holy Scriptures.

He then expounds the positive religious value of his probable speculations: when we know that there are a multitude of such earths inhabited, we shall be less apt to admire what this world calls great and we shall adore God's Providence and wisdom manifested all over the universe 'to the confusion of those who would have the earth and all things formed by the shuffling concourse of Atoms, or to be without beginning'.

This clearly shows that even such a thorough-going mechanist as Huygens believed in an underlying creative plan for the world.

Huygens states optimistically that the arguments of Galileo, Gassendi and Kepler have brought all astronomers to the Copernican side, except those who have 'their Faith at another man's disposal, *and so for fear of Galileo's fate dare not own it*'. (The words about Galileo have been added by the English translator who evidently thought that it was not clear enough at whom this thrust was levelled.) Now, if the earth has a moon, and Jupiter and Saturn have theirs, it is probable that in other things they agree also, and that all other planets have inhabitants like the earth. This 'must be our method in this treatise, wherein from the nature and circumstances of that planet which we see before our eyes, we may guess at those that are farther distant from us'. Plants and animals are so exactly adapted to some design that it would be 'an absurdity to think that they are haphazardly jumbled together by chance motion of little particles' (a thrust at Epicurus' atomism and at Descartes). From the fact that plants and animals in America resemble those in the Old World, Huygens infers that it is probable that such a correspondence exists also between the living beings in other planets and the earth. So there must be creatures there, perhaps not men, but also endued with reason. Huygens cannot believe that all beauty and furniture of the planets would be enjoyed by the Creator alone and not by some beings for whose sake they have been made. He thinks that the inhabitants of the planets live a settled life in a society and 'not a wandering Scythian way of life', a remarkable conclusion for a man who has just ridiculed philosophers who have an anthropomorphic idea of God and now puts forward an anthropomorphic view of non-human extra-terrestrial beings. The inhabitants of the planets must be as good architects as ourselves: 'Who are we? a company of mean fellows living in a little corner of the world, upon a ball 10,000 times less than Jupiter and Saturn. And yet we forsooth must be the only skilful people at building?'

The doctrine of the plurality of worlds inevitably changes the belief that all things are for the sake of man into a belief that every particular world is for the sake of its inhabitants. In Huygens' opinion, saying 'that "all things are made for the use of man" cannot mean that they are just for us to stare or peep through a telescope at them'. 'The greatest part of God's Creation, that innumerable multitude of stars, is placed out of the reach of any man's eye,' so that they seem not to belong to us. 'Holy Scripture only speaks about this Earth and man, the lord of them, but as for Worlds in the sky it is wholly silent.'

Huygens cannot believe that the wise Creator has put all His animals and plants on this spot alone and has left all the other worlds destitute of inhabitants who might adore Him and that all those prodigious bodies were made only to twinkle to be studied by some of us 'poor fellows'.

We saw Huygens boldly apply analogical reasoning. For the sceptical Robert Boyle things were not so easy. He had recognized the probability of other worlds, fixed stars with planetary systems, visible or beyond our sight. He points out that the animals of the New World are different from those of Europe; the more so those of other planets must be different. He is not sure even whether the same fundamental laws of motion are valid in those systems.

What common attitudes to (a) Scripture and (b) nature may be detected in the writings of Wilkins and of Huygens?

Both are united in their views on the sufficiency of Scripture. For neither was it true that all truth was embedded in the Biblical text. Consequently, one cannot deny the existence of other inhabited worlds merely because Scripture does not happen to mention them. Equally, neither man was prepared to contradict the positive assertions of Scripture. Turning to their attitudes to nature, Wilkins and Huygens are both clear that it was a creation by God; Wilkins himself distinguished between the plurality of worlds (acceptable) and the plurality of universes, which could have implied a plurality of creators and formerly led to polytheism—a conclusion he found totally unacceptable. The insistence by Huygens that the universe must reflect the glory of God is a characteristic and widely held notion of his own times and later.

We have now seen that the Christian orthodoxy of scientists sometimes found the plurality of worlds a highly edifying doctrine. On the purely theological side we might mention Richard Baxter (1667), a Nonconformist minister of the Restoration period. He recognized that plurality is no revealed truth, and believed that as practically all parts of this our world are populated by some animal, it seems almost certain that this will be the case with the more important parts of Creation, which will have inhabitants proportionate to their grandeur.

In the nineteenth century Thomas Chalmers, the Scottish theologian who was to play such an important role in the Disruption, made a plea for a plurality of worlds, and Thomas Dick, a Scottish schoolmaster, even found it warranted by Holy Scripture. In his *Sidereal Heavens* (1840), a work 'illustrative of the character of the Deity, and of an infinity of worlds', he recognized that the Scriptures are chiefly of a moral nature, and not intended to teach us the principles of physical sciences. Yet, the doctrine of the plurality of worlds 'is embodied in many passages of the sacred writings'. A direct statement would be Hebrews 11: 3 (and 1: 2) which he chose as motto of the work: 'the worlds were framed by the word of God'.

There were also some astronomers of note like Bode and the Herschels who believed that other celestial bodies are inhabited. The well-known historian of science, William Whewell, who in his Bridgewater Treatise held the pluralistic view, in 1853 defended the uniqueness of the earth and its inhabitants. On the other hand, the physicist David Brewster (1854) accepted the plurality of worlds on philosophical and theological grounds.

Theologically and scientifically the discussion raised the problem whether it is sensible to speculate on matters about which scanty or even no data at all are available. At any rate the discussion helped to make man think about his modest place in the universe and it stimulated research in physical astronomy. Nevertheless, it often verged on science fiction. But what would science be without a fiction element? Analogical reasoning may be a bane or a blessing to science, but it is indispensable.

Further Reading

Giorgio de Santillana, *The Crime of Galileo*, Chicago, 1955; Heinemann Education, Mercury paperback edition, London, 1961.

F. R. Johnson, *Astronomical Thought in Renaissance England*, 1937, reprinted Octagon Press, 1968.

E. Zinner, *Entstehung und Ausbreitung der Coppernicanischen Lehre*, Verlag Max Mencke, Erlangen, 1943.

J. L. E. Dreyer, *A History of Astronomy from Thales to Kepler*, Dover Publications no 353.

R. Hooykaas, *The Reception of Copernicanism in the Netherlands*, North Holland Publishing Co. and Royal Netherlands Academy of Sciences, Amsterdam, 1974.

Acknowledgements

Grateful acknowledgement is made to the following sources for material used in this unit:

Text

The Officers and Council of the Royal Astronomical Society for Copernicus, *De Revolutionibus* (trans. J. F. Dobson) in *Occasional Notes*.

Illustrations

Figures 1, 6 and 8: Mansell Collection; *Figures 2, 5, 9 and 10*: Ronan Picture Library; *Figure 2 (bottom)*: Royal Astronomical Society; *Figures 3, 11 and 15*: Royal College of Physicians; *Figure 4*: Staatsbibliothek, Berlin: Preussischer Kulturbesitz; *Figure 7*: The Warden and Fellows of Wadham College, Oxford. Photo: B. J. Harris; *Figures 12, 13, 14 and 16*: British Museum.

Unit 3
Galileo and the Church

Prepared by David C. Goodman for the Course Team

Contents

1 Introduction

Of all the issues concerning science and religion none has excited more controversy or aroused more passion than Galileo's conflict with the Roman Catholic Church. A century ago the German historian Karl von Gebler spoke of the 'malice and fierce partisanship' which had been responsible for the fables and exaggerations in current historical accounts of the Galileo affair. In his *Galileo Galilei and the Roman Curia* (first published in German in 1876, shortly after translated into English, and still much admired), von Gebler attempted to write an objective biography of Galileo, based on all the evidence that was available to him, including access to the Vatican files. He wanted a true history freed from the generally accepted myths:

> According to these legends, Galileo languishes during the trial in the prisons of the Inquisition; when brought before his judges, he proudly defends the doctrine of the double motion (of the earth); he is then seized by the executioners of the Holy Office, and subjected to the horrors of torture; but even then—as heroic fable demands—he for a long time remains steadfast; under pain beyond endurance he promises obedience, that is, the recantation of the Copernican system. As soon as his torn and dislocated limbs permit, he is dragged before the large assembly of the Congregation, and there, kneeling in the penitential shirt, with fierce rage in his heart, he utters the desired recantation. As he rises he is no longer able to master his indignation, and fiercely stamping with his foot, he utters the famous words: 'E pur si muove!' ('And yet it moves!') He is, therefore, thrown into the dark dungeons of the dreaded tribunal, where his eyes are put out![1]

Von Gebler showed that hardly a single detail of the above account was true, and gradually these distortions disappeared from the history books. But the study of the causes of Galileo's dispute with the Church has continued to test the impartiality of twentieth-century historians. In 1938 F. Sherwood Taylor, a historian of science, estimated that around two hundred books and papers had been written on the career of Galileo. Nevertheless he felt there was still room for his *Galileo and the Freedom of Thought*. He complained that existing works had failed to bring out that Galileo was fighting for the progress of science, and that his involvement in theological controversy was an unhappy accident. He regretted that Catholic historians had 'unfairly decried Galileo's achievement, over-praised the attitude of his antagonists, and minimized the paramount importance of the principles for which Galileo strove'.

More recently Giorgio de Santillana's *The Crime of Galileo* has appeared. The author, Professor of the History and Philosophy of Science at the Massachusetts Institute of Technology, states that the Galileo affair is far from dead, and continues:

> As far as I can make out, the unsolved state of the puzzle is due to this: that the 'free-thinkers' are only too glad to put the Roman Church as a whole under accusation in the affair, while inside the ecclesiastic hierarchy powerful corporate interests are ready to accept the terrain chosen by the attackers rather than have some of their long-deceased members stand under the spotlight of history. Thus, they are willing to involve the Church in this quarrel, with the inevitable consequences that they have to resort to laborious smoke screens, devious implications, and all sorts of unfair tactics.[2]

An appeal to the Church to allow the spotlight of history to fall more fully on the Galileo case has been made by Arthur Koestler.[3] Koestler said uncertainty would

---

[1] Karl von Gebler, *Galileo Galilei and the Roman Curia* (trans. Mrs George Sturge), London, 1879, pp. 249–50.

[2] Giorgio de Santillana, *The Crime of Galileo*, Chicago, 1955; Heinemann Education, Mercury paperback edition, London, 1961, p. xii.

[3] Arthur Koestler, *The Sleepwalkers: A History of Man's Changing Vision of the Universe*, Hutchinson, 1961.

remain on important aspects of the case until 'the complete Vatican file is at last made accessible to scholars' (p. 601). It is interesting that Koestler confesses to the reader in advance that his own account is not free from bias. He admits that he has an aversion towards Galileo, chiefly because of his behaviour towards Kepler. On the other hand Koestler was not inclined to feel sympathy for the Inquisition in Rome:

> Among the earliest and most vivid impressions of History was the wholesale roasting alive of heretics by the Spanish inquisition, which could hardly inspire tender feelings towards that establishment.[1]

Liking neither Galileo nor the ecclesiastical authorities, Koestler's narrative was based on the belief that the conflict between the two sides need not have occurred at all: 'I believe the idea that Galileo's trial was a kind of Greek tragedy, a show-down between "blind faith" and "enlightened reason", to be naively erroneous'.[2]

It should be clear from what has already been said that a biography of Galileo highlights some of the most basic problems facing the historian: the separation of fact and myth, the ideal of an objective appraisal free from the distortions of prejudice, the need for reinterpretation as fresh evidence becomes available. As we will see, acute problems also arise on the authenticity of historical documents. You should bear all of this in mind in this unit, particularly when reading the section on Galileo's trial.

In this unit we will above all be concerned with explaining why Galileo's support of the Copernican system should have brought ecclesiastical opposition, culminating in a trial before the Inquisition. After studying this unit, the associated reading material, and the related television programme 2, you should be able to:

1 Describe the principal features of Aristotelian cosmology and physics.
2 Discuss the significance of Galileo's telescopic observations for Aristotelian, Ptolemaic and Copernican theories.
3 State the reasons for Galileo's belief in the Copernican system.
4 Give an account of Galileo's opinions on science and religion and contrast this with the policy of the Roman Catholic Church at the time.
5 Discuss some of the most important arguments put forward in Galileo's *Dialogue*.
6 Distinguish between the parties who were hostile and those who were sympathetic to Galileo.

2 Galileo's Criticism of Contemporary Cosmology

2.1 His early opposition to the professors

Galileo Galilei was born in 1564 in Pisa, then part of the Grand Duchy of Tuscany, ruled by Cosimo de Medici. Following a Tuscan custom of naming the first-born, the Christian name Galileo was a repetition of the surname.

His father Vincenzio Galilei was an accomplished musician and composer with some knowledge of mathematics. He sent his son to the University of Pisa to read medicine, but Galileo disliked the course which reverentially followed the texts of the ancients (Aristotle, Galen). Already Galileo began to challenge the truth of respected classical doctrine. He became interested in mathematics, and this was to remain with him throughout his life.

Figure 1   *Galileo, from the painting by Sustermann* (*Crown Copyright: Science Museum*).

---

[1] Koestler, *op. cit.*, p. 425.

[2] Koestler, *op. cit.*, p. 426.

At the age of twenty-five, through the influence of a nobleman, Guidobaldo dal Monte, Galileo was appointed professor of mathematics at the University of Pisa. According to the well-known story it was during his brief tenure of this post that Galileo dropped balls of lead and iron from the Leaning Tower in the presence of university professors, to falsify Aristotle's theory that heavy bodies fell to the ground with a velocity in direct proportion to their weight. However, this story is now discredited, as is the legend that Galileo's discovery of the law of pendulum motion was the result of watching the oscillations of the great candelabra in the cathedral of Pisa. (It is now known that at this time no candelabrum had yet been introduced into the cathedral.)

Perhaps because of the university's resentment of Galileo's opposition to the established doctrines of Galen and Aristotle, on which the curriculum was based, the young man left Tuscany in 1592 to take up the vacant chair of mathematics at the University of Padua in the Republic of Venice. Here Galileo produced his most important work, the kinematics that would transform physical science. However, this remained unpublished until his old age. It seems that his lectures on astronomy were conventional, that is based on Ptolemy. But in private he was already a convinced Copernican. This is clear from a letter to Kepler, then a teacher of mathematics at a school in Graz, thanking him for the gift of the *Mysterium Cosmographicum*, a book in which Kepler expressed his belief in the Copernican system. Galileo wrote:

> I promise to read your book in tranquility, certain to find the most admirable things in it, and this I shall do the more gladly as I adopted the teaching of Copernicus many years ago, and his point of view enables me to explain many phenomena of nature which certainly remain inexplicable according to the more current hypotheses. I have written many arguments in support of him and in refutation of the opposite view—which however, so far I have not dared to bring into the public light, frightened by the fate of Copernicus himself, our teacher, who though he acquired immortal fame with some, is yet to an infinite multitude of others (for such is the number of fools) an object of ridicule and derision. I would certainly dare to publish my reflections at once if more people like you existed; as they don't, I shall refrain from doing so.[1]

So Galileo, like Copernicus, feared not the opposition of the Church, but the ridicule of the ignorant. Kepler replied urging Galileo to publish the proof he claimed to have for the Copernican system. But apart from a book on a mathematical instrument invented by him, Galileo published nothing until 1610, when his little book *The Starry Messenger* appeared. As a result Galileo became widely known.

2.2   The Aristotelian universe

If we are to appreciate the impact of *The Starry Messenger*, and of Galileo's subsequent works, we have to understand the prevailing thought concerning the cosmos. At the time of Galileo the structure of the universe was still conceived in accordance with the ideas of one of the most influential of all of the ancient philosophers, Aristotle (384–322 B.C.). In his comprehensive works Aristotle constructed an entire system of philosophy which became widely accepted in ancient Greece. His cosmology profoundly influenced the thought of the Arabs and was later dominant in medieval Western Christendom.[2]

In the medieval universities of Western Europe Aristotle was an established authority whose writings formed a basis for instruction in logic and natural philosophy. Similarly the Church used Aristotle's texts to reinforce the Scriptures. This had not been without resistance, for Christians already possessed a body of ideas about the universe in the Scriptures, which conflicted with the reasoning of the pagan Greek. For example, Aristotle had written that the universe had existed from eternity,

[1] Quoted in Koestler, *op. cit.*, p. 356.

[2] See R. Hooykaas, *Religion and the Rise of Modern Science*, Scottish Academic Press/Chatto and Windus (SET BOOK), pp. 1–13, 31–2.

which opposed the Judaeo-Christian belief in a creation. In the early years of the thirteenth century ecclesiastical authorities had issued decrees prohibiting the teaching of Aristotle's philosophy, for example at the University of Paris in 1210. However, this trend was soon reversed as a result of the synthesis of Albertus Magnus (*c.* 1200–1280) and Thomas Aquinas (1225–74). They showed how Aristotelian philosophy could be joined to Christian theology, a marriage which has existed ever since in Catholicism. While the Church accommodated Aristotle's philosophy, it rejected the interpretation that Aristotle had said all there was to say on the natural world. Therefore Aristotle's theories could be rejected without incurring ecclesiastical opposition. During the Middle Ages natural philosophers frequently criticized Aristotle's physics, above all his treatment of moving bodies on earth, and this continued with Galileo.

According to Aristotle the universe was a sphere of limited size, beyond which nothing existed. At the centre of the universe lay the earth, at rest. This was the proper, natural place for the earth, since he said heavy bodies in motion, such as detached pieces of earth, fall down towards the centre. The earth was surrounded by a series of concentric spheres, first by the sphere of water, then by spheres of air and fire. The flames of a fire lit on earth rose up, indicating that the natural place for fire was above the earth, and that the motion appropriate to fire was in a straight line upwards, so differing in nature from the natural falling motion of terrestrial objects.

The sphere of fire was surrounded by a succession of seven solid, crystalline spheres which carried the moon, Mercury, Venus, the sun, Mars, Jupiter and Saturn respectively. The next sphere, the outermost in the universe, contained the fixed stars. This sphere rotated and so communicated a circular motion to the inner spheres carrying the planets. The illustration below, taken from a sixteenth-century work, shows why Aristotle's universe has been compared to the skins of an onion.

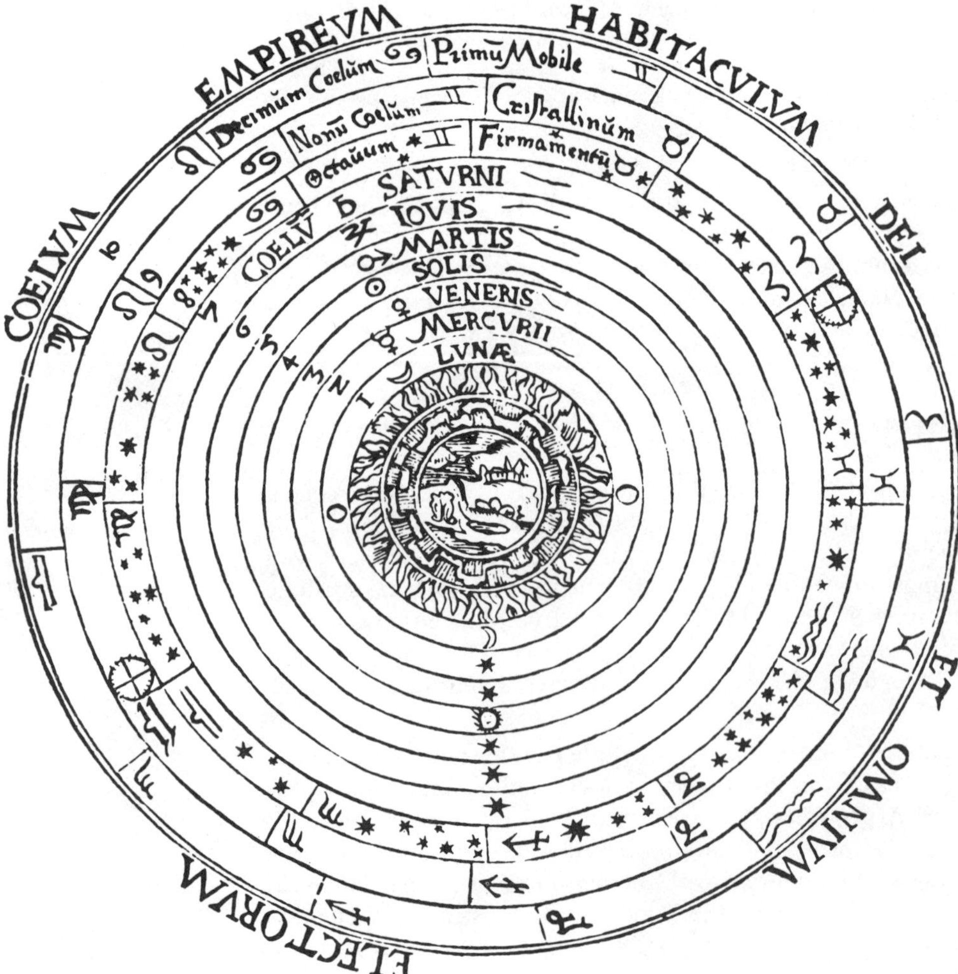

Figure 2   Geocentric system of the universe showing Aristotle's four elements, earth, water, air and fire surrounded by the spheres of the planets and the fixed stars, from Petrus Apianus, Cosmographia per Gemma Phrysius restituta, *Antwerp, 1539* (Ronan Picture Library).

The sphere of the moon was a fundamental boundary in the Aristotelian universe. Two distinct sets of laws operated in the universe: one set for the region beneath the moon, the other for the moon and the spheres beyond. Aristotle regarded the sublunary region as one of imperfection, characterized by generation and decay. There was no permanence; movements from place to place occurred in straight lines, and so were destined not to last, since circular motion alone was everlasting. The celestial region, on the other hand, from the moon upwards, was unalterable and indestructible. Aristotle said that no one in recorded history had ever witnessed a change in the heavens. The celestial regions had no imperfections. No earth, water, air or fire, the constituent elements of the terrestrial world, were to be found there. Instead a special fifth element, the quintessence, was alone fitted for the substance of the heavens. In accordance with their everlasting nature, the moon and bodies beyond moved uniformly in a never-ending circle. The idea originated from theological considerations. Aristotle asserted that the ultimate cause of motion in the universe was God, who set the concentric spheres in motion by moving the outermost sphere of stars. However, God Himself did not move. According to Aristotle the unmoved Mover caused motion not by any physical contact with the stellar sphere, but spirtually by being an object of love and desire. For a celestial sphere to be capable of love it must be endowed with a psyche or soul, and in fact Aristotle not only gave a psyche to each of his spheres, but likened the heavenly bodies to animals. Each sphere therefore craves to be like the unmoved Mover. The spheres come closest to this ideal, not by being unmoved, but by acquiring the perfection of an everlasting circular motion.

Could Aristotle's view of the universe accommodate the postulates of the Copernican system?

So long as Aristotle's cosmology prevailed it presented an impenetrable barrier to the Copernican system. Copernicus had postulated that the earth was moving in orbit, that is that the earth was a planet. Aristotle's cosmology, on the other hand, denied motion to the earth and separated it from the planets. Moreover Copernicus stated that the earth moved naturally in a circle. As we have just seen, Aristotle's cosmology allowed only rectilinear motion to the earth, that is to displaced parts of the earth.

2.3 The evidence of the telescope

In 1609 news reached Galileo that a new instrument had been invented in the Netherlands which made distant objects appear close to the observer. Galileo tried various combinations of lenses to achieve this effect, and he soon succeeded with a concave and a convex lens placed at the opposite ends of a lead tube; the concave lens formed the eyepiece. His third telescope made objects appear 1,000 times larger and over thirty times closer than when seen with the naked eye.

Galileo remarked that the value of the new instrument at sea or on land was obvious. But he was more concerned with astronomical observations, and apparently for the first time he turned a telescope to the skies. The astonishing results were published by Galileo with the title *The Starry Messenger*.

2.3.1 The surface of the moon

The surface of the moon appeared pitted with craters. There were mountain peaks lit by the sun's light, while large areas of the moon remained in darkness. Galileo compared the effect to a sunrise in a mountainous region of the earth; he likened a part of the moon's surface to Bohemia.

Galileo thought incorrectly that the darker areas of the moon were seas. He calculated that some of the moon's mountains were about four miles high, a very good estimate.

Figure 3 Galileo's drawings of the surface of the moon (Crown Copyright: Science Museum).

How well did Galileo's observations of the moon agree with the Aristotelian view of the universe?

The observations conflicted with the fundamental Aristotelian belief that the moon and all heavenly bodies beyond possessed a state of perfection which was unknown on earth. Galileo had found that the moon's surface, far from being perfectly smooth and polished, was full of irregularities; indeed its unevenness seemed even to exceed that of the earth.

This was powerful evidence against the Aristotelian universe. It reinforced the earlier observations of the Danish astronomer Tycho Brahe, who, before the invention of the telescope, had already weakened the idea of celestial perfection in another way. Brahe observed the new star which appeared in the constellation Cassiopeia in 1572 and established that it had been generated beyond the planetary regions. Similarly in 1577 he showed that a comet moved beyond the sun. This discovery falsified the Aristotelian theory that comets were meteorological events occurring in the sublunary region and showed that crystalline spheres in the heavens could not exist, since the comet of 1577 had passed through them. Brahe's observations had begun to undermine the belief that no changes ever occurred in the heavens.

2.3.2 A multitude of stars

Galileo also turned his telescope to the stars, and as a result was able to dismiss an objection which Brahe had raised against Copernicus. Galileo found that the stars, unlike the moon, were hardly enlarged at all through the telescope. He explained correctly that the naked eye tends to overestimate the size of stars, judging their magnitude by their brightness. Adventitious rays of light, spreading out from the star, reached the eye and gave the impression that the star was bigger than it really was. This brightness surrounding stars could be observed in their twinkling. The telescope excluded many of these rays which reached the naked eye over a wide angle.

Brahe had said that if the stars were really as distant as Copernicus had stipulated for them not to show stellar parallax, their dimensions would become incredible. However, Brahe's argument had been based on the assumption that stellar magnitudes were proportional to their brightness as seen by the naked eye. In showing that the stars through the telescope appeared not as enlarged discs but as points, Galileo disposed of this obstacle to the Copernican system.

In another respect Galileo's telescopic observations of the stars added to the plausibility of the Copernican universe. Galileo was overwhelmed by the hundreds of stars, which for the first time became visible with the telescope. The Milky Way, which for naked-eye astronomy had been a pale glow of uncertain nature, was resolved by the telescope into a myriad of individual stars. Similarly, nebulae previously regarded as reflected light were now shown to consist of groups of small stars. The revelation of a multitude of hitherto unknown stars at very great distances from the earth supported not only Copernicus' own statements which portrayed the universe as more vast than generally considered, but also those followers of Copernicus, such as Thomas Digges, who had speculated on an infinite universe. Galileo himself, however, like Kepler, could not go so far as to accept that the universe was infinite.

2.3.3 The Medicean stars

But Galileo's *Starry Messenger* contained an even stronger argument in support of Copernicus. Of all his telescopic observations he regarded his discovery of the satellites of Jupiter as the most pregnant. On 7 January 1610 Galileo noticed three very bright objects close to Jupiter. At first he regarded them as fixed stars lying in a straight line. But on successive nights he found the following relative positions of Jupiter and the three bodies, eventually observing a fourth:

(i)	East	✪ ✪ • ✪	West
(ii)	East	• ✪ ✪ ✪	West
(iii)	(Clouds prevented observation on the third night)		
(iv)	East	✪ ✪ •	West
(v)	East	✪ ✪ •	West
(vi)	East	✪ ✪ • ✪	West
(vii)	East	✪ • ✪ ✪	West

How could these observations be explained? (Ignore the slight alteration in the alignment of the bodies in the final observation.)

One possible explanation, which Galileo at first entertained, was that the change in the relative positions of Jupiter and the small bright bodies was due to the motion of Jupiter. For example, the pair of observations (i) and (ii) can be explained by supposing an eastward movement of Jupiter. However, the relative positions observed on subsequent nights showed this was unacceptable—the motion of Jupiter would have to be highly irregular. Above all the variation in size of the bright objects confirmed that their distance from Jupiter was changing. Galileo denied that this variation in size could be an illusion produced by observing through the atmosphere of the earth since the size of Jupiter remained the same, while that of the neighbouring bodies altered. The variation in the number of bodies represented on the various nights was due to their being hidden from view by Jupiter.

Galileo explained that there were four satellites revolving about Jupiter. Moreover, Galileo added in his first published expression of sympathy for the Copernican theory, Jupiter and its satellites together revolved about the sun, the centre of the universe. Jupiter's satellites had provided him with a miniature version of the Copernican universe:

> Here we have a fine and elegant argument for quieting the doubts of those who, while accepting with tranquil mind the revolutions of the planets about the sun in the Copernican system, are mightily disturbed to have the moon alone revolve about the earth and accompany it in an annual rotation about the sun. Some have believed that this structure of the universe should be rejected as impossible. But now we have not just one planet rotating about another while both run through a great orbit around the sun; our own eyes show us four stars which wander around Jupiter as does the moon around the earth, while all together trace out a grand revolution about the sun in the space of twelve years.

So the earth was not unique in having a satellite. Furthermore, just as Jupiter's satellites kept up with the orbiting Jupiter, so it was conceivable that the earth could orbit the sun without the moon being left behind.

Galileo christened Jupiter's satellites the 'Medicean stars' after the ruling family of Tuscany. Indeed he had dedicated *The Starry Messenger* to Cosimo II de Medici, once his pupil and now the Grand Duke. Perhaps because of homesickness Galileo had sought to win favour with the Tuscan ruler. A few months after the publication of his first astronomical work Galileo was appointed chief mathematician and philosopher to the Grand Duke and head of mathematics at the University of Pisa. It is worth noting that there was a significant difference in the political situation at Pisa and Padua. Unlike Pisa, and the rest of his native Tuscany, Padua was free of interference from Rome. This was because the Venetian Republic had recently succeeded in its struggle with the Pope to rule its own affairs. It has often been pointed out that had Galileo stayed in Padua, he would never have been brought before the Inquisition.

In *The Starry Messenger* Galileo had given no evidence for Copernicus' postulate that the earth was a planet, revolving annually about the sun. He did, however, promise proof of this in future work which would contain a comprehensive treatment of the universe:

> We shall prove the earth to be a wandering body surpassing the moon in splendour, and not the sink of all dull refuse of the universe: this we shall support by an infinitude of arguments drawn from nature.

2.3.4 The phases of Venus

A few months after the announcement of his first telescopic observations Galileo communicated to astronomers (including Kepler) his momentous discovery of the phases of Venus. This was nothing less than the first falsification of the established Ptolemaic system of astronomy. Copernicus had predicted that a crucial test, involving the appearance of Venus, would decide between his heliocentric theory and Ptolemy's geocentric astronomy. For if Venus moved around the sun it should exhibit a sequence of phases to an observer on earth, varying like the moon from a circle to a crescent. On the other hand if Venus revolved about the earth as centre, as Ptolemaic astronomy maintained, the observer would never see more than a thin crescent of Venus. Before the telescope was available, Venus was too small for the naked eye to detect phases. But Galileo was able to see for the first time that Venus appeared to change its shape, like the moon. He also observed the marked variation in the apparent size of Venus, showing that its distance from the earth was altering. This was easily explained on the Copernican (and Tychonic) system, since Venus at various points in its orbit might be on the same or the different side of the sun in relation to the earth.

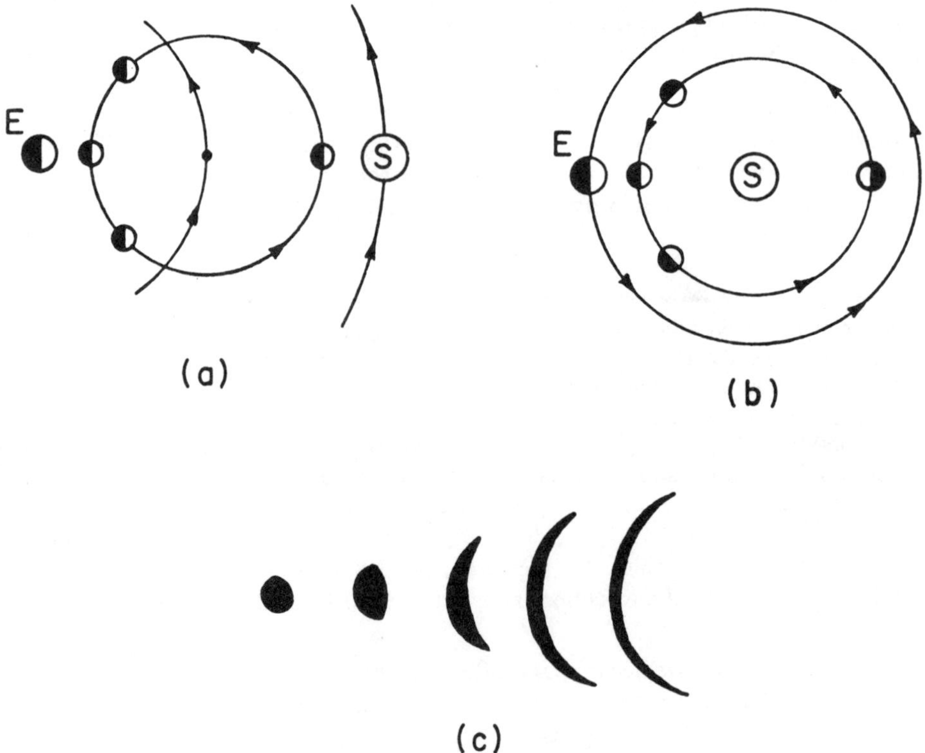

Figure 4 The phases of Venus in (a) the Ptolemaic system, (b) the Copernican system, and (c) as observed with a low-power telescope. In (a) an observer on the earth should never see more than a thin crescent of the lighted face. In (b) he should see almost the whole face of Venus illuminated just before or after Venus crosses behind the sun. This almost circular silhouette of Venus when it first becomes visible as an evening star is drawn from observations with a low-power telescope on the left of diagram (c). The successive observations drawn on the right show how Venus wanes and simultaneously increases in size as its orbital motion brings it closer to the earth (T. S. Kuhn, The Copernican Revolution, *Harvard University Press/Oxford University Press, 1957, Fig. 44, p. 223).*

Galileo's observations on Venus destroyed the dominant position of Ptolemaic astronomy, but did not establish the truth of the Copernican system. For the appearance of Venus was also consistent with Tycho Brahe's theory which made all the planets, except the earth, revolve around the sun.

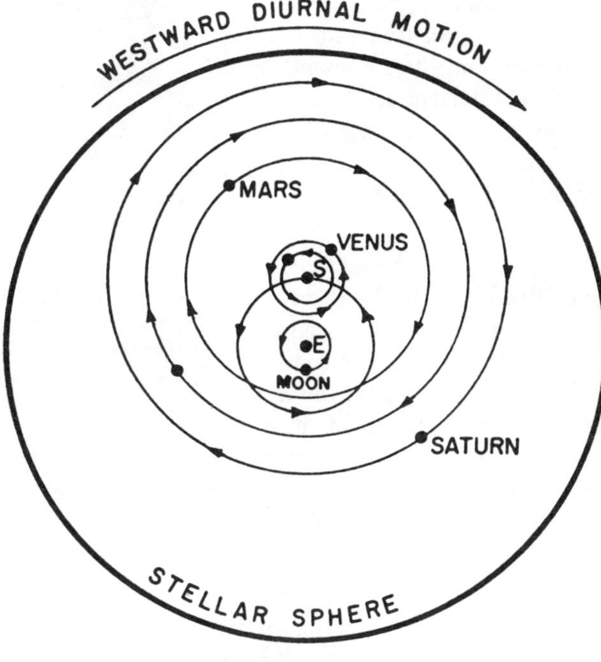

Figure 5 The Tychonic System. The earth is once again at the centre of a rotating stellar sphere, and the moon and sun move in their old Ptolemaic orbits. The other planets are, however, fixed on epicycles whose common centre is the sun (T. S. Kuhn, The Copernican Revolution, *Harvard University Press/Oxford University Press, 1957, Fig. 37, p. 202).*

2.3.5 The reception of Galileo's observations

Hundreds of copies of *The Starry Messenger* were soon printed. Kepler received the work with enthusiasm and had a second edition printed at Frankfurt. Others were less enthusiastic. Libri, who taught philosophy at Pisa, refused to look through the telescope. There were allegations that Galileo's observations were all illusions created by the telescope.

This was not entirely absurd, since it was well known that a single lens could distort, so a combination of two lens could be regarded with even more suspicion. Galileo offered a large reward to anyone who could make an instrument which would display bright objects moving around Jupiter but not about any other celestial body.

A more serious difficulty arose in confirming the observations, since Galileo's telescope was superior to many that were being constructed. In Rome Father Clarius, the Jesuit authority on astronomical matters, had been sceptical but he was soon to confirm that the new stars and satellites of Jupiter existed. The Jesuit order attached great importance to intellectual pursuits. Their studies included mathematics and astronomy. Certain Jesuits acquired an expertise in these subjects and were consulted by the Church when necessary. They even discussed Galileo's observations in China, where they had gone as missionaries.

Galileo visited Rome in 1611 and was ceremoniously received by Clarius and other Jesuits of the Roman College. When the head of the Roman College, Cardinal Robert Bellarmine, asked the Jesuit mathematicians for their opinion of Galileo's discoveries, he was told they were confirmed. Galileo had a cordial interview with Pope Paul V.

During this visit Galileo was elected a member of the Accademia dei Lincei, a society dedicated to the pursuit of learning especially of natural philosophy. Headed by a nobleman Federico Cesi, who was Galileo's friend, this academy was a predecessor of later scientific societies such as the Royal Society. The Accademia dei Lincei provided Galileo with an important means for communicating his ideas, and also sponsored his later publications.

In all, Galileo's visit to Rome had been a great success. His standing with the Church seemed high. Yet Cardinal Bellarmine took the precaution to write to the

Chief Inquisitor at Padua to ask if Galileo's name had been mentioned in a recent case (this involved proceedings against Cremonini, an Aristotelian philosopher who had aroused Church opposition for denying the immortality of the soul). Galileo had no involvement in the case, but the Cardinal's action was significant. Bellarmine was the leading theologian of the Catholic Church and the guardian of orthodoxy. He had dealt with the parties which had challenged papal supremacy: the Lutherans, Gallicans and the Republic of Venice. He had been one of the Inquisitors who had tried Giordano Bruno, for heretical views on transubstantiation and the Immaculate Conception. Bruno was sent to the stake in 1600. Bellarmine was determined to avoid another such case. Apart from his theological heresies Bruno had been a supporter of Copernicus. Bellarmine was alert to the possibilities of further trouble, and in the excitement surrounding Galileo's visit, he wanted to be sure that the astronomical developments would bring no turmoil.

Figure 6 *Robert Bellarmine (1542–1621), engraved by Francis Villamena after School of Cornelis Cort (Mansell Collection).*

At this stage Galileo had less to fear from the Church than from the university professors, whose reputation and professional occupations depended on the survival of Aristotelian cosmology and physics. No wonder that some had simply refused to look through the telescope. The Aristotelian opposition to Galileo was organized by Lodovico delle Colombe, who does not seem to have taught at a university, though he strongly defended the Aristotelian beliefs which he shared with the academics. Little is known about this man, but it is clear from contemporary correspondence that he was Galileo's principal enemy. A letter to Galileo in 1611 from his friend Ludovico Cigoli, an artist, referred in mocking terms to the 'pigeon league', a phrase often to be used by Galileo's supporters for the faction led by Colombe, whose name is the Italian word for 'dove'.

Colombe had recently written a book entitled *Contro il Moto della Terra* (Against the Motion of the Earth). This made no mention of Galileo, but the author took the important step of opposing Copernicanism with quotations from the Scriptures to support the belief that the earth was at rest in the centre of the universe. Colombe was prepared if necessary to create a religious scandal in the interests of Aristotelianism. The evidence shows that he was behind the outbursts of certain friars who were soon to assail Galileo.

2.3.6   *Letters on Sunspots*

On his return from Rome Galileo became involved in two quarrels. His dispute with Colombe on floating bodies culminated in Galileo's *Discourse on Floating Bodies*, which appeared in 1612 and which attacked Aristotelian physics. Soon after this he had an unpleasant argument with Father Christopher Scheiner, a Jesuit astronomer, each claiming priority for the discovery of sunspots. In fact they were observed on occasions many centuries before—sunspots are visible to the naked eye.

Scheiner, writing under a pseudonym, had described his observations in letters to Mark Welser, an Augsburg merchant and an enthusiast of natural philosophy. Blemishes on the face of the sun were unacceptable to Scheiner who, as an Aristotelian, believed in the perfection of the heavens. So he attributed them to small planets obstructing our view of the sun as they passed close to it.

Galileo's letters in reply were sent to Mark Welser, and published in 1613 as *Letters on Sunspots* by the Accademia dei Lincei. He argued that the sun, like the moon, was not free from imperfections. Sunspots changed their shape irregularly, which he said argued against Scheiner's assertion that they were stars. Because of their transient nature Galileo compared the sunspots to clouds, which also could be immense, were of a shifting nature, and were capable of obstructing sunlight. He suggested a model to imitate sunspots by the use of bitumen on a red-hot iron plate. A black cloud of smoke was produced which varied in shape. He could not decide if the spots were actually located on the sun's surface, or a very small distance from it. Galileo also observed the rotation of the sunspots and correctly attributed it to the axial rotation of the sun.

Unlike *The Starry Messenger*, which had been written in Latin, Galileo's *Letters on Sunspots* were in Italian and so could be read by a far larger number in Italy. The *Letters* are noteworthy not only for their further evidence against the Aristotelian cosmos. For the first time Galileo openly asserted that the Copernican system was the only true account of the heavens.

The following passages occur in the *Letters on Sunspots*. Read them carefully and then do the exercises which follow (you may need to refer back to Unit 2, to refresh your mind on eccentrics, deferents, equants and epicycles).

Criticizing Scheiner's astronomy, Galileo wrote:

1 ... he continues to adhere to eccentrics, deferents, equants, epicycles, and the like as if they were real, actual, and distinct things. These, however are merely assumed by mathematical astronomers in order to facilitate their calculations. They are not retained by philosophical astronomers who, going beyond the demand that they somehow save the appearances, seek to investigate the true constitution of the universe— the most important and most admirable problem that there is. For such a constitution exists; it is unique, true, real, and could not possibly be otherwise; and the greatness and nobility of this problem entitle it to be placed foremost among all questions capable of theoretical solution.

2 ... although it may be vain to seek to determine the true substance of the sunspots, still it does not follow that we cannot know some properties of them, such as their location, motion, shape, size, opacity, mutability, generation, and dissolution. These in turn may become the means by which we shall be able to philosophize better about other and more controversial qualities of natural substances. And finally by elevating us to the ultimate end of our labours, which is the love of the divine Artificer, this will keep us steadfast in the hope that we shall learn every other truth in Him, the source of all light and verity.

Write a paragraph comparing Galileo's approach to astronomy described in the first passage, with that of Copernicus (Unit 2). Compare the religious sentiments expressed by Galileo in the second passage with text 1 in *Anthology I*, an extract from Kepler.

1 This is a clear statement of the two fundamentally different approaches which had divided astronomers since antiquity. Galileo could not follow the tradition which restricted astronomy to mathematical techniques describing the movements of celestial bodies and to the calculations which predicted their future positions. This is precisely what Ptolemaic astronomy had become, and in the spurious preface to Copernicus' *De Revolutionibus*, this was the way the Protestant clergyman Osiander had said astronomers ought to proceed. On this view astronomers were free to select any of a great number of alternative mathematical techniques, according to their simplicity in use, elegance, and above all ability to describe the celestial movements. The one consideration that did not arise was the physical truth of the theory employed. You will recall from the previous unit that Osiander had presented Copernicus' system as an ingenious calculating device with no pretensions to portray the actual state of the universe, and that he had taken this step in order to avoid offending theologians through a theory which contradicted scriptural passages.

Copernicus' own attitude was nothing of this kind. He regarded his system as a true account of the world, and the first passage shows that this was also Galileo's approach to astronomy. Galileo wanted to discover the actual structure of the universe. It was a question of finding a unique solution, the true theory. Galileo deeply believed that the universe was mathematically constructed, but he could not be satisfied with mathematical theories which did not aspire to give a real account of the cosmos. It was just because of his firm adherence to this interpretation of astronomy, and of the whole of natural philosophy, that Galileo was to come into conflict with contemporary theologians.

2 Galileo was a devout Catholic. No one has doubted the genuineness of his concern for the Church, or the integrity of his faith. This short passage shows that religious beliefs were a motivation for his scientific activities, whose ultimate goal was knowledge and love of the divine creator, through the discovery of the laws by which God had ordered the natural world. The same sentiments were much more pronounced in Kepler, a Protestant, as the text in *Anthology I* referred to shows. Indeed the entirety of Kepler's momentous astronomical work was inspired by a fervent desire

to discover the divine plan of the universe. The same motive lay behind the scientific research of may natural philosophers of the seventeenth century. The persistence of this fundamental connection between religion and science in later centuries will be discussed in subsequent units.

Galileo's reference to God as the source of light is also worth some comment. The analogy of light had been employed in the past by St Augustine (354–430) to describe the process by which divine truths were communicated to the human mind: it was a process of illumination. This idea taken literally had given optics, the study of light, a special importance since the Middle Ages. An influential theory which regarded optics as the key to understanding the universe supposed a role of fundamental importance for the radiation of light from a central source. Kepler described the generation of the universe in these terms. Light and power radiated from the central source of the universe, which for Kepler represented God the Father: this was the sun. It was this idea, rather than any scientific proof, which convinced Kepler that the Copernican heliocentric universe was correct.

The *Letters on Sunspots* brought to a close Galileo's period of astronomical observation. His exciting discoveries with the telescope had strengthened the position of the Copernican system, but it is important to stress they had by no means established its truth. The telescope had undermined Aristotelian cosmology, falsified the Ptolemaic system, but gave no proof for the Copernican theory. Galileo had promised to prove that the earth was an orbiting planet, but so far he had not done so. There was no scientific evidence against Tycho Brahe's version of the geocentric universe, which had the further attraction of conforming to the orthodox view, sanctioned by common sense, scriptural authority and academic tradition, that the earth was at rest. This theory was still a serious alternative to Copernicanism in the middle of the seventeenth century. Galileo gave the theory less attention than it deserved and was never able to eliminate it. Galileo had in fact committed himself to the Copernican system before he could prove that it was uniquely true. The illustration below, taken from a Tychonic exposition of astronomy by the Jesuit Father Riccioli in 1651, shows the Tychonic system weighing more heavily than that of Copernicus.

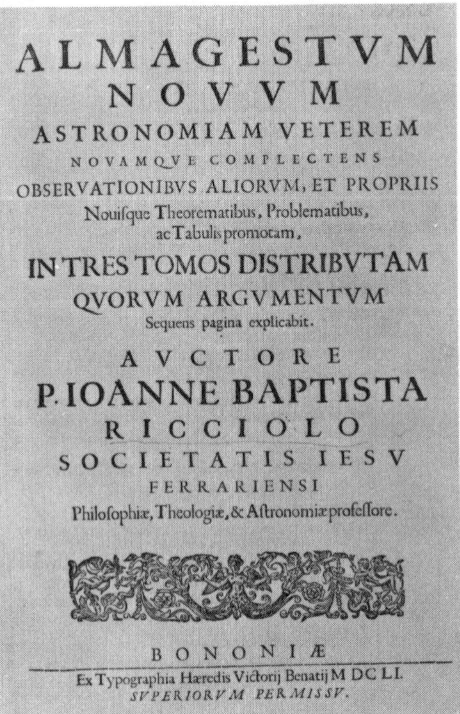

Figure 7 (a) and (b)   Frontispiece and title page of Giovanni Battista Riccioli, Almagestum Novum, astroniam veterem novamque complectens, *1651, volume 1 (British Museum).*

3 Galileo and Theology

3.1 Clerical censures

Towards the end of 1611 Galileo received the first inkling that theologians might come out against him. A letter from his friend Cigoli in Rome brought the news that a group hostile to Galileo had met at the home of an archbishop to decide how best to deal with the unorthodox opinions which included the motion of the earth. One proposal was to get a clergyman to denounce the opinions from the pulpit, but for the while no action was taken. The archbishop in question was the Archbishop of Florence. However, it is not at all clear that he disapproved of Galileo's ideas.

The first recorded criticism of Galileo by a priest did not come until the end of 1612 and occurred not in a public sermon but in a private conversation. Father Niccolo Lorini, a Dominican monk, had remarked that the opinions of 'Ipernicus—or whatever his name is' supported by Galileo were contrary to the Holy Scriptures. Galileo quickly wrote to the old monk and received a courteous reply saying that he cared little about the discussion and had only spoken a few words to show he was still alive. Galileo was soon joking about a critic who could not even get Copernicus' name right.

A more serious incident occurred on 21 December 1616, and this time Galileo was far from joking. In the Dominican church of Santa Maria Novella in Florence, Father Thomas Caccini preached a fiery sermon against Galileo, the Copernican system, and all mathematicians, declaring them enemies of Christianity. He referred to the miracle described in the tenth chapter of the Book of Joshua, one of the Biblical passages which seemed to assert that the sun was in motion:

'Sun, stand thou still at Gibeon,
and thou moon in the valley of Ajalon'.
And the sun stood still and the moon stayed,
until the nation took vengeance on their enemies.

He is also supposed to have made a pun on the Biblical text (Acts 1:11): 'Ye men of Galilee, why stand ye gazing up into heaven?'

Caccini's outburst brought an apology from the Master-General of the Dominican Order in Rome, who wrote to Galileo that it was the misfortune of his office to have to answer for the stupidities of some of his tens of thousands of brethren. Caccini had been disciplined before for his unethical behaviour. Evidence of his connection with the 'pigeon league' in this affair comes in the form of a letter from his brother, a curial official, rebuking him in the strongest terms for allowing Colombe and the pigeon league to drive him on to behave so disgracefully. It is difficult to avoid seeing Caccini as anything but an arch-villain in the Galileo affair.

3.2 *Letter to the Grand Duchess Christina*

But the real source of Galileo's dispute with the Church was a discussion at the dinner table of the Grand Duke Cosimo II in December 1613. The guests included Boscaglia, a professor of philosophy at the University of Pisa, and Benedetto Castelli, a Benedictine monk who had been a pupil of Galileo in Padua. A convinced Copernican, he was an ardent disciple of Galileo. He had just been appointed to the chair of mathematics at Pisa with specific instructions not to teach the motion of the earth. The conversation at dinner turned on the university, the telescope and astronomical observations. The Grand Duchess Christina, the mother of the Grand Duke, asked questions about Jupiter's satellites and assurance was given that they really existed. At the end of the meal, and in Castelli's absence, Boscaglia confided in the Grand Duchess that while all of Galileo's observations were confirmed, his statement that the earth moved was incredible, chiefly because it contradicted the Holy Scriptures. Castelli, who was on his way out of the palace, was overtaken by a servant of the Grand Duchess with a request to return. In the discussion which resumed the Grand Duchess quoted scriptural passages against Castelli's Galilean opinions. Castelli used theological arguments which persuaded the Grand Duke. The Grand Duchess continued to dispute, but Castelli construed this as merely a pretext for hearing his replies. Throughout Boscaglia remained silent.

Castelli described all that had taken place in a letter to Galileo. A week later Galileo sent a very long letter in reply expounding his ideas on the relationship between science and religion. Castelli circulated copies of this letter amongst his friends. A copy fell into the hands of old Father Lorini, the Dominican priest who had uttered a few words against 'Ipernicus'. The contents shocked him and he at once notified the Holy Office, the Inquisition in Rome, enclosing a version of the letter which was not entirely faithful to the original. Lorini was angry and deplored the meddling of a layman into theological matters and the expression of heretical opinions. The Bishop of Fiesole called for the jailing of Copernicus and had to be informed that this man had been dead for a number of years.

Galileo was anxious that the authentic version of his letter should reach Rome, so he recovered the original from Castelli and sent it to his friend, Piero Dini, an archbishop, asking for it to be shown if possible to Cardinal Ballarmine. Galileo added that he would compose an enlarged version of his letter to Castelli which had been written in a hurry. This completed statement of Galileo's opinions on the subject of science and religion appeared under the title *Letter to the Grand Duchess Christina*, addressed to the dowager who had questioned his disciple. (It circulated as a manuscript and was not published until over twenty years later.)

The *Letter to the Grand Duchess Christina* is reproduced in full in *Anthology I* (text 3). Read this together with Hooykaas' *Religion and the Rise of Modern Science*, pp. 124–30, and then write short answers to the following.

1 **What were the reasons given by Galileo for writing the *Letter to the Grand Duchess Christina*?**
2 **What were the sources regarded by Galileo as giving a true account of things, and what was his basic belief in the relation of these sources?**
3 **How did Galileo explain why conflict had arisen between natural philosophy and certain scriptural passages?**
4 **What authority did Galileo attach to the pronouncements of the Fathers of the Church?**
5 **Does Galileo argue for the separation of scientific research from all religious matters?**
6 **Can you find a passage in which Galileo argues for the Copernican system in a way different from before?**
7 **Compare Galileo's views on science and religion with those of the Puritan clergyman John Wilkins, whose Copernican *Discourse Concerning a New Planet* of 1640 supported the earth's planetary status. Text 6 of *Anthology I* contains an extract from this work.**

1 Galileo was writing to clear himself of the charges of heresy levelled against him by his opponents in the academic world. The academic arguments of the 'pigeon league' did not matter to him, since he was convinced that the future would bring their collapse. But he could not tolerate the accusations of heresy. In addition he hoped his *Letter* would guide the Church to make a wise pronouncement on the Copernican system.

2 For Galileo the Bible and natural philosophy gave the truth about the world. The Holy Scriptures were divinely inspired and therefore contained the truth revealed as the Word of God. The study of natural phenomena, or what Galileo called 'the open book of Nature', was another avenue to truth by way of observation and reason. This path also proceeded from God, since the natural world was created in accordance with His divine commands. God had given us senses and the ability to reason, and with these aids it was possible for natural philosophers to discover truths about the created world.

Since the book of Nature and the Bible both derived from God, Galileo asserted that it was impossible for the one to contradict the other. Is so doing he followed the example of Augustine who had reconciled Christian faith and Greek science with the same doctrine of consistent truths. As you have seen, Galileo's *Letter* is full of quotations from Augustine.

3   Again following Augustine, Galileo explained how contradictions between natural philosphy and the Holy Sciptures might arise. This could occur if the statements made by natural philosophers were not true, because of errors of observation or argument. However, the difficulty still arose with the rigorously demonstrated truths of natural philosophy, in which class Galileo included the Copernican beliefs in a mobile earth and a stationary sun. In cases such as this Galileo said it was necessary to go beyond the surface of the scriptural passages, the words of which were not to be taken literally. Once the true meaning of the words was found, Galileo was certain the contradiction would vanish, since contradiction was impossible for the reasons already given.

Galileo illustrated this doctrine with references to Biblical passages, which taken literally were contradictory and even heretical, as in the texts which attributed hands and feet to God and the human imperfections of anger and ignorance. Galileo said these words were written by the scribes for the common people. Similarly he said that whenever the Bible touched on difficult physical matters, it was written in language which the simple people of the time could understand. That was why, as Thomas Aquinas pointed out, the Bible referred to the emptiness of celestial space, while learned men knew it was filled with air (*sic*).

The state of motion or rest of the earth and sun were also described in the Bible in accordance with the common opinion of the time. Galileo argued that this use of words was necessary to secure the assent of the people to the principal articles of faith. If the Bible had said the earth moved and the sun stood still, the population at large would have believed neither statement, and doubt would have been cast on other scriptural texts. Galileo said even in the far less primitive population of his own day, not one in a thousand would deny the common sense view that the earth was at rest; nor were ordinary folk in a position to understand the argument of the opposite side.

Again quoting Augustine, Galileo argued that the Bible was concerned above all with the salvation of souls. If the sacred scribes had intended to teach astronomy they would have written much more on the subject; but in fact the planets were hardly mentioned. He repeated the witty remark, said to have been made by Cardinal Baronius, which seemed to sum the matter up admirably. The purpose of the Holy Ghost in the Scriptures was to teach us 'how to go to heaven, not how the heavens go'.

4   In reply to the view that an agreed opinion by the holy Fathers should be binding, even on physical propositions, Galileo maintained that this could only be accepted if the Fathers had debated the issues thoroughly, if they had come to an agreement and if they had condemned an opposite opinion. None of these conditions was met according to Galileo, chiefly because in the past the question of the earth's mobility or sun's rest had not been controversial questions of concern to the Fathers of the Church. In any case the Catholic doctrine defined by the Council of Trent in its ruling on this, called for the acceptance of the holy Fathers' interpretation of the Scriptures only on matters of faith or morals. Galileo said the behaviour of the sun and earth were not in this category.

5   Not in the sense that science and religion have nothing to do with one another, since Galileo's belief in a Creator was one of the motivations for his scientific work. However, Galileo was insisting on the freedom to pursue scientific research to arrive at conclusions on physical problems independently, unhampered by the constraints of existing opinion, especially of theological opinion. He said the road of natural philosophy must not be closed as if everything was already discovered. He could not accept the position of those theologians who in good faith demanded that natural philosophers should search for a fallacy in any of their conclusions which contradicted the sacred texts, the repository of truth. Galileo remarked that this was to ask the impossible: 'For this would amount to commanding that they must not see what they see and must not understand what they know, and that in searching they must find the opposite of what they actually encounter.' No created being, not even he said the absolute power of the Pope, could declare the phenomena of nature to be

other than what they were in reality. The prohibition of the Copernican system would be tantamount to the prohibition of astronomy, since men would be asked not to recognize the phases of Venus.

So Galileo was led to delineate the roles of natural philosophy and the Bible as sources of information. In all matters concerning the supernatural or on problems that were inaccessible to human reason (he gave as an example the question whether the stars were animate) Galileo said the Scriptures must be rigorously adhered to. On the other hand in all that concerns natural phenomena it was necessary to employ observations and reasoning instead of scriptural texts. No physical conclusion that is satisfactorily demonstrated should be subordinated to a Biblical passage; on the contrary it was up to the theologians to show by a judicious interpretation that the Bible did not contradict physical demonstrations. In this way natural philosophy would indicate the true meaning of ambiguities in the Scriptures.

It is clear that Galileo was anxious not only to secure freedom for natural philosophy, but also to avoid jeopardizing the position of the Church and the Bible by the ecclesiastical condemnation of scientifically established conclusions. Centuries before, Augustine had warned Christians not to make positive statements on matters which were not understood: the nature of animals, fruits and stones, the properties of the earth, or the distances of the stars. For if such hazardous opinions turned out to be false, outsiders would associate these errors with Christian doctrine and would give no credence to fundamental principles like the resurrection of the dead or the belief in the Kingdom of Heaven. Galileo agreed with all of this, pointing out that this situation now threatened to arise with the possible condemnation of the statement of the earth's motion. The desire of the Church to bring outsiders into the fold would be thwarted if prelates were moved by men, whose real motive was the preservation of vested academic interests by any means including the flashing of the Church sword.

6   As you will have read, at the end of the letter Galileo argued cleverly that the passage in the Book of Joshua, which seemed to imply that the sun had an orbital motion was really more consistent with the Copernican system than with Ptolemaic astronomy. In the course of this Galileo gave a revealing description of the sun, which he believed was rotating on its axis stationary at the centre of the universe:

> . . . if we consider the mobility of the sun, and the fact that it is the font of light which (as I shall conclusively prove) illuminates not only the moon and the earth but all the other planets, which are inherently dark, then I believe that it will not be entirely unphilosophical to say that the sun, as the chief minister of Nature and in a certain sense the heart and soul of the universe, infuses by its own rotation not only light but also motion into other bodies which surround it. And just as if the motion of the heart should cease in an animal, all other motions of its members would also cease, so if the rotation of the sun were to stop, the rotations of all the planets would stop too.

This passage shows that Galileo's belief in the Copernican system was at least partly inspired by a type of sun worship, which in various forms has occupied an important place in the thought of primitive societies, the ancient Egyptians, etc. Copernicus similarly had spoken of the sun as 'ruler of the universe' and could think of no more appropriate location for the 'throne' than the centre of the universe. Closely related to the medieval Neo-platonic tradition which attached a religious significance to light (which we mentioned earlier), this metaphysical belief in the special qualities of the sun convinced Copernicus and his followers, Kepler and Galileo, of the correctness of the heliocentric universe before they could provide the necessary proof. In his *Letter* Galileo gave no evidence that would raise the Copernican system to the level of a demonstrated truth, despite his belief that this was the case.

7   Like Galileo's *Letter*, the extract from Wilkins's *Discourse Concerning a New Planet* distinguished between the procedures to be adopted in acquiring truths on spiritual matters and in natural philosophy. Recourse to the Scriptures was necessary for the former, but in natural philosophy it was preposterous to begin with the opinions of others. So like Galileo, Wilkins was in effect trying to clear the way for scientific research. (Wilkins was himself a natural philosopher who helped to organize scientific meetings in Oxford when he was Warden of Wadham College.)

Unlike Galileo, who wrote that the sacred scribes had concealed their knowledge of Copernican astronomy in order to accommodate the beliefs of the common people, Wilkins asserted that Moses, Job and the other scribes were ignorant of such philosophical truths. Joshua's command to the sun to stop moving was a reflection of his ignorance of astronomy.

As Professor Hooykaas points out, while Galileo looked to the Bible for veiled statements of the Copernican system, Wilkins insisted that this modern scientific opinion was not to be found there.

3.3 Cardinal Bellarmine's reactions

Nothing came of old Father Lorini's protest to the Inquisition over Galileo's views on science and the Scriptures. The authorities in Rome could find nothing heretical in them, but the intrusion of a layman in theological matters was not welcomed. Galileo's enlightened interpretations were not endorsed by the Catholic Church until the end of the nineteenth century. Galileo was secretly under the scrutiny of the Church.

Figure 8 Title page of Paolo Antonio Foscarini, Lettera . . . sopra l'Opinione de' Pittagorici, e del Copernico della mobilita della terra, e stabilita del sole e del nuovo Pittagorico sistemo del Mondo, 1615 (British Museum).

Cardinal Bellarmine's comment after Galileo's *Letter to Castelli* had been read to him was that there was no question of banning Copernicus' book, only perhaps adding a little to it to stress that the work did not pretend to give a true account of the universe. He said that with the same precaution Galileo could discuss the theory whenever he chose. The Cardinal was soon induced to make a more detailed statement. A Carmelite friar, Paolo Foscarini, had just sent Bellarmine a copy of his book defending the Copernican system, and asserting like Galileo, that the system was physically true. This book seems to have encouraged Galileo to press his views on the theologians. Bellarmine was not encouraged, he wrote to Foscarini:

I have gladly read the letter in Italian and the essay in Latin that Your Reverence has sent me, and I thank you for both, confessing that they are filled with ingenuity and learning. But since you ask for my opinion, I shall give it to you briefly, as you have little time for reading and I for writing.

First. I say that it appears to me that Your Reverence and Sig. Galileo did prudently to content yourselves with speaking hypothetically and not positively, as I have always believed Copernicus did. For to say that assuming the earth moves and the sun stands still saves all the appearances better than eccentrics and epicycles is to speak well. This has no danger in it, and it suffices for mathematicians. But to wish to affirm that the sun is really fixed in the centre of the heavens and merely turns upon itself without travelling from east to west, and that the earth is situated in the third sphere and revolves very swiftly around the sun, is a very dangerous thing, not only by irritating all the theologians and scholastic philosophers, but also by injuring our holy faith and making the sacred Scripture false. For your Reverence has indeed demonstrated many ways of expounding the Bible, but you have not applied them specifically, and doubtless you would have had a great deal of difficulty if you had tried to explain all the passages that you yourself had cited.

Second. I say that, as you know, the Council of Trent would prohibit expounding the Bible contrary to the common agreement of the holy Fathers. And if Your Reverence would read not only all their works but the commentaries of modern writers on Genesis, Psalms, Ecclesiastes, and Joshua, you would find that all agree in expounding literally that the sun is in the heavens and travels swiftly around the earth, while the earth is far from the heavens and remains motionless in the centre of the world. Now consider whether, in all prudence, the Church could support the giving to Scripture of a sense contrary to the holy Fathers and all the Greek and Latin expositors. Nor may it be replied that this is not a matter of faith, since if it is not so with regard to the subject matter, it is with regard to those who have spoken. Thus that man would be just as much a heretic who denied that Abraham had two sons and Jacob twelve, as one who denied the virgin birth of Christ, for both are declared by the Holy Ghost through the mouths of the prophets and apostles.

Third. I say that if there were a true demonstration that the sun was in the centre of the universe and the earth in the third sphere, and that the sun did not go around the earth but the earth went around the sun, then it would be necessary to use careful consideration in explaining the Scriptures that seemed contrary, and we should rather have to say that we do not understand them than to say that something is false which had been proven. But I do not think there is any such demonstration, since none has been shown to me. To demonstrate that the appearances are saved by assuming the sun at the centre and the earth in the heavens is not the same thing as to demonstrate that in fact the sun is in the centre and the earth in the heavens. I believe that the first demonstration may exist, but I have very grave doubts about the second; and in case of doubt one may not abandon the Holy Scriptures as expounded by the holy Fathers. I add that the words 'The sun also riseth and the sun goeth down, and hasteth to the place where he ariseth' (Ecclesiastes 1: 5) were written by Solomon, who not only spoke by divine inspiration, but was a man wise above all others of all created things, which wisdom he had from God; so it is not very likely that he would affirm something that was contrary to demonstrated truth, or truth that might be demonstrated. And if you tell me that Solomon spoke according to the appearances, and that it seems to us that the sun goes round when the earth turns, as it seems to one aboard ship that the beach moves away, I shall answer thus. Anyone who departs from the beach, though to him it appears that the beach moves away, yet knows that this is an error and corrects it, seeing clearly that the ship moves and not the beach; but as to the sun and earth, no sage has needed to correct the error, since he clearly experiences that the earth stands still and that his eye is not deceived when it judges the sun to move, just as he is likewise not deceived when it judges that the moon and the stars move. And that is enough for the present.[1]

Write a short paragraph comparing and contrasting Cardinal Bellarmine's opinions on astronomy and the Scriptures with those of Galileo.

[1] From *Discoveries and Opinions of Galileo*, translated with introduction and notes by Stillman Drake, Doubleday Anchor paperback, 1957, pp. 162–4.

Bellarmine, the chief spokesman for the ecclesiastical position on matters of controversy, had outlined a policy that was not totally opposed to Galileo's stand. Like Galileo he argued that if a demonstrated truth in astronomy contradicted a scriptural passage, this could only mean that the true meaning of the latter was not yet known. However, since he denied that the Copernican postulates were demonstrated truths, Bellarmine could not accept Galileo's assertion that this applied to scriptural passages relating to the sun and earth. In addition Bellarmine declared, in contrast to Galileo, that the interpretations of the Fathers were to be accepted as matters of faith whatever the subject of their discussion. Just as the Protestant theologian Osiander had advised Copernicus, Cardinal Bellarmine said Galileo would be wise to discuss the Copernican system as a calculating device and not as a physically true account. This would avoid theological controversy. As we have seen, Galileo was no more prepared than Copernicus to accept this weakening of astronomy. The Cardinal had made one telling point. He remarked that there was in any case no conclusive proof for the Copernican system.

3.4 The Decree of 1616

Galileo's stand was uncompromising and he decided to go to Rome to put his case in person. Against the advice of the Tuscan ambassador in Rome, Galileo arrived there on 11 December 1615 determined to win a debate. He discussed the Copernican theory on many occasions, but although he was able to destroy the arguments put forward in support of Ptolemaic astronomy and Aristotelian cosmology, he failed to convince his audience that the Copernican system was physically true.

Finally Pope Paul V intervened with a request for an official statement on the motion of the earth and the stability of the sun.

A report was issued by the theological experts of the Congregation of the Index. On 24 February 1616 the theologians announced their decisions on the controversial matters they had been summoned to consider. The proposition that the sun was stationary at the centre of the universe was declared 'foolish and absurd, philosophically and formally heretical, in as much as it expressly contradicts the doctrine of the Holy Scripture in many passages, both in their literal meaning and according to the general interpretation of the Fathers and Doctors'. The other proposition submitted for consideration: that the earth was neither at the centre of the universe nor immovable, but moved as a whole and also with a diurnal motion, was unanimously pronounced deserving of 'the same censure in philosophy and, as regards theological truth, to be at least erroneous in faith'.

A week later Foscarini's Copernican treatise was put on the Index of prohibited books. The printer concerned was imprisoned and Foscarini himself soon died in mysterious circumstances. Copernicus' *De Revolutionibus* on the other hand was not condemned outright. It was to be 'corrected' by a few changes which Galileo himself regarded as slight: removing the word 'star' from the descriptions of the earth and taking a few lines from the Preface which maintained that the theory was in harmony with the Scriptures. In this modified version Copernicus' work could be read again in 1620. The Church had shown that while it would not tolerate Copernican works which called for a reinterpretation of the Scriptures, the Copernican system could be discussed hypothetically. The followers of Copernicus were not declared heretics, since this would have required a solemn proclamation from the Pope or General Council of the Church. Since this did not occur the theological decisions against the Copernican postulates did not mean that they were incompatible with the Catholic faith.

What of Galileo himself? None of his books was placed on the Index, nor was he mentioned by name. Yet he had been at the centre of the controversy. Following an order from the Pope, Cardinal Bellarmine summoned Galileo for interview. An account of this meeting was later issued by Bellarmine at Galileo's request. It reads as follows:

We, Roberto Cardinal Bellarmino, having heard that it is calumniously reported that Signor Galileo Galilei has in our hand abjured and has also been punished with salutary penance, and being requested to state the truth as to this, declare the said Signor Galileo has not abjured, either in our hand, or the hand of any other person here in Rome, or anywhere else, so far as we know, any opinion held by him; neither has any salutary penance been imposed on him; but that only the declaration made by the Holy Father and published by the Sacred Congregation of the Index has been notified to him, wherein it is set forth that the doctrine attributed to Copernicus, that the Earth moves around the Sun and that the Sun is stationary in the centre of the world and does not move from east to west, is contrary to the Holy Scriptures and therefore cannot be defended or held. In witness whereof we have written and subscribed these presents with our hand this twenty-sixth day of May, 1616.

This certificate was carefully kept by Galileo. In a later section you will be asked to refer back to this text in order to compare it with another document which gave a different version of what had happened at the interviews. As we will see, the discrepancy between the two documents was to be of decisive importance at his trial.

Galileo took some satisfaction from the Church's decision not to condemn Copernicanism outright. Caccini's continued caustic outbursts had not been listened to. He was pleased his name had been cleared from malicious gossip. But he had not won his battle and left Rome dejected.

Two years later Galileo sent the Archduke of Austria an explanation of the causes of the tides. This was not simply an attempt to investigate a difficult phenomenon; for the tides were regarded by Galileo as nothing less than the best proof of the earth's motion, the vital missing link in the Copernican debate. In view of the 1616 decree he had to write cautiously. So with ironical language he told the Archduke that the work was just 'a poetical conceit, or a dream'. He developed his theory in a subsequent work of greater importance, and so we will deal with the tidal theory in a later section.

4 Dialogue Concerning the Two Principal Systems of the World

4.1 Galileo's discussions with the new Pope

Galileo's spirits rose in 1623 with the election of a new Pope, Urban VIII, formerly Maffeo Barberini, a member of the Accademia dei Lincei who had written a poem in honour of Galileo's telescopic discoveries. There was some reaon to hope that a Pope with interests in natural philosophy and who had accepted the dedication in the *Assayer*, a philosophical work by Galileo (published in 1623), might show favour to Copernican astronomy, and perhaps even reverse the recent decrees. In 1624 Galileo returned to Rome and had several audiences with the new Pope. It soon became clear that the Church's attitude was not going to change just because Bellarmine and the former Pope were now dead. Galileo was free only to present the Copernican system as a convenient hypothesis.

Galileo now began to work on a treatise of the world that he had promised fifteen years before. After several delays he completed the book in 1629, and arrived in Rome in May 1630 with his manuscript *On the Flux and Reflux of the Sea*, seeking official approval for the printing. The Pope repeated his demand that the treatment be hypothetical. He also requested that the title of the work should be changed to *Dialogue Concerning the Two Principal Systems of the World* (that is the Ptolemaic and the Copernican). The importance which Galileo attached to tidal phenomena is nowhere more clearly expressed than in the original title to his treatise. However, the Pope did not want the Copernican discussion to be presented under a title which gave prominence to a phenomenon which Galileo believed he could explain demonstratively.

Further, the Pope insisted that Galileo's discussion of the tides should be weakened with an assertion that the cause considered was only one of the ways God might have employed to bring about the tides. Accordingly Galileo wrote a preface and altered the conclusion of his work. After hesitations by the official responsible for granting the license, Riccardi, who was not at all certain that Galileo had complied with the Pope's directions, the *Dialogue* was published in Florence in 1632.

4.2 Form and content of the *Dialogue*

Written in the vernacular, and so capable of being widely read in Italy, the book was in the form of Socratic dialogues, partly, Galileo said, because this would allow interesting digressions. This style also gave Galileo the opportunity of honouring deceased friends who would play two of the three characters involved in the discussions. One was Giovanni Sagredo, a Venetian nobleman who had scientific discussions with Galileo during his Paduan period. The other friend, Filippo Salviati, may have studied with Galileo at Padua. He came from a Florentine banking family. At his villa near Florence he and Galileo carried out observations on sunspots.

Against a Venetian setting, Galileo developed a discussion over four days between Sagredo, Salviati and a third character, Simplicio. The extracts below are taken from the 'Second Day'.[1] (The *Dialogue* is the subject of a radio programme in course A201.)[2]

Read them carefully and then answer the questions which follow.

1

SALV. As the strongest reason of all is adduced that of heavy bodies, which, falling down from on high, go by a straight and vertical line to the surface of the earth. This is considered an irrefutable argument for the earth being motionless. For if it made the diurnal rotation, a tower from whose top a rock was let fall, being carried by the whirling of the earth, would travel many hundreds of yards to the east in the time the rock would consume in its fall, and the rock ought to strike the earth that distance away from the base of the tower. This effect they support with another experiment, which is to drop a lead ball from the top of the mast of a boat at rest, noting the place where it hits, which is close to the foot of the mast; but if the same ball is dropped from the same place when the boat is moving, it will strike at that distance from the foot of the mast which the boat will have run during the time of fall of the lead, and for no other reason than that the natural movement of the ball when set free is in a straight line toward the center of the earth. This argument is fortified with the experiment of a projectile sent a very great distance upward; this might be a ball shot from a cannon aimed perpendicular to the horizon. In its flight and return this consumes so much time that in our latitude the cannon and we would be carried together many miles eastward by the earth, so that the ball, falling, could never come back near the gun, but would fall as far to the west as the earth had run on ahead.

They add moreover the third and very effective experiment of shooting a cannon ball point-blank to the east, and then another one with equal charge at the same elevation to the west; the shot toward the west ought to range a great deal farther out than the other one to the east. For when the ball goes toward the west, and the cannon, carried by the earth, goes east, the ball ought to strike the earth at a distance from the cannon equal to the sum of the two motions, one made by itself to the west, and the other by the gun, carried by the earth, toward the east. On the other hand, from the trip made by the ball shot toward the east it would be necessary to subtract that which was made by the cannon following it. Suppose, for example, that the journey made by the ball in itself was five miles and that the earth in that latitude travelled three miles during the flight of the ball in the shot toward the west, the ball would fall to earth eight miles distant from the gun— that is, its own five toward the west and the gun's three to the east. But the shot toward

---

[1] From Galileo Galilei, *Dialogue Concerning the Two Chief World Systems—Ptolemaic and Copernican*, translated by Stillman Drake, University of California Press paperback, 1962.

[2] The Open University (1972) A201 *Renaissance and Reformation*, Radio 16, *Authority and Freedom: Galileo's Dialogue*.

the east would range no further than two miles, which is all that remains after subtracting
from the five of the shot the three of the gun's motion toward the same place. Now
experiment shows the shots to fall equally; therefore the cannon is motionless, and
consequently the earth is, too. Not only this, but shots to the south or north likewise
confirm the stability of the earth; for they would never hit the mark that one had aimed
at, but would always slant toward the west because of the travel that would be made
toward the east by the target, carried by the earth while the ball was in the air. And not
merely shots along the meridians, but even those made to the east or west would not
range truly; for the easterly shots would carry high and the westerly low whenever they
were aimed point blank. Since the shots in both directions take the path of a tangent—
that is, a line parallel to the horizon—and the horizon is always falling away to the east
and rising in the west if the diurnal motion belongs to the earth (which is why the eastern
stars appear to rise and the western stars to set), it follows that the target to the east would
drop away under the shot, wherefore the shot would range high, and the rising of the
western target would make the shot to the west low. Hence in no direction would shooting
ever be accurate; and since experience is contrary to this, it must be said that the earth is
immovable.

SIMP. Oh, these are excellent arguments, to which it will be impossible to find a valid
answer.

SALV. Perhaps they are new to you?

SIMP. Yes, indeed, and now I see with how many elegant experiments nature graciously
wishes to aid us in coming to the recognition of the truth. Oh, how well one truth accords
with another, and how all cooperate to make themselves indomitable!

SAGR. What a shame there were no cannons in Aristotle's time! With them he would
indeed have battered down ignorance, and spoken without the least hesitation concerning
the universe.

SALV. It suits me very well that these arguments are new to you, for now you will not
remain of the same opinion as most Peripatetics, who believe that anyone who departs
from Aristotle's doctrine must therefore have failed to understand his proofs. But you
will certainly see further novelties; you will hear the followers of the new system producing
observations, experiments, and arguments against it more forcible than those adduced by
Aristotle and Ptolemy and the other opponents of the same conclusions. Thus you will
become assured that it is not through ignorance or inexperience that they have learned to
adhere to such opinions.

2

SIMP. I tell you, as I have told you on other occasions, that the greatest master there has
been for teaching the recognition of sophisms, paralogisms, and other fallacies is Aristotle,
who in this particular can never be mistaken.

SAGR. You are only angry that Aristotle cannot speak; yet I tell you that if Aristotle were
here he would either be convinced by us or he would pick our arguments to pieces and
persuade us with better ones. For look: Did not you yourself, upon hearing the experiments with cannons described, understand and admire them, and confess them more
conclusive than Aristotle's arguments? Yet I do not hear Salviati, who put them forward
and who has surely examined them and explored them minutely, confess himself persuaded
by them, nor even by others of still greater force which he intimates that he is about to
deliver to us. And I do not know upon what basis you accuse Nature of having been for
many ages in her second childhood, having forgotten how to produce any reflective
thinkers except those who make themselves slaves of Aristotle and have to think with his
brain and see with his eyes.

But let us hear the rest of the arguments favorable to his opinion so that we may proceed
with their testing, refining them in the crucible and weighing them in the assayer's balance.

SALV. Before going further I must tell Sagredo that I act the part of Copernicus in our
arguments and wear his mask. As to the internal effects upon me of the arguments which
I produce in his favor, I want you to be guided not by what I say when we are in the heat
of acting out our play, but after I have put off the costume, for perhaps then you shall find
me different from what you saw of me on the stage.

Now let us proceed. Ptolemy and his followers produce another experiment like that
of the projectiles, and it pertains to things which, separated from the earth, remain in the
air a long time, such as clouds and birds in flight. Since of these it cannot be said that they
are carried by the earth, as they do not adhere to it, it does not seem possible that they
could keep up with its swiftness; rather, it ought to look to us as if they were being moved
very rapidly westward. If we, carried by the earth, pass along our parallel (which is at

least sixteen thousand miles long) in twenty-four hours, how could the birds keep up on such a course? Whereas we see them fly east just as much as west or any other direction, without any detectable difference.

Besides this, if, when we travel on horseback, we feel the air strike rather strongly upon our faces, then what an east wind should we not perpetually feel when being borne in such a rapid course against the air! Yet no such effect is felt.

Here is another very ingenious argument taken from certain experiences. Circular motion has the property of casting off, scattering, and driving away from its center the parts of the moving body, whenever the motion is not sufficiently slow or the parts not too solidly attached together. If, for example, we should very rapidly spin one of those great treadmills with which massive weights are moved by one or more men walking within them (such as huge stones used in mangles, or barges being dragged across the land from one waterway to another), then if the parts of this rapidly turned wheel were not very solidly joined, it would all come apart. Or, if many rocks or other heavy materials were strongly attached to its external surface, they would not be able to resist the impetus, and it would scatter them with great force to various places far from the wheel, and accordingly from its center. If, then, the earth were to be moved with so much greater a velocity, what weight, what tenacity of lime or mortar would hold rocks, buildings, and whole cities so that they would not be hurled into the sky by such precipitous whirling? And men and beasts, none of which are attached to the earth; how would they resist such an impetus? Whereas on the contrary, we see these and the much less resistant pebbles, sand, and leaves reposing quietly upon the earth, and even falling back upon it with very slow motion.

Here, Simplicio, are the very potent arguments taken, so to speak, from terrestrial things. There remain those of the other kind; that is, those with relation to celestial appearances, which arguments tend still more to show that the earth is in the center of the universe, and consequently deprive it of the annual motion around that center as attributed to it by Copernicus. These being of rather a different nature, they can be brought forth after we have judged the strength of those already propounded.

SAGR. Well, what do you say, Simplicio? Does it seem to you that Salviati understands and knows how to explain the Ptolemaic and the Aristotelian arguments? Do you think any Peripatetic understands the Copernican proofs so well?

SIMP. Had I not formed from previous arguments such a high opinion of Salviati's soundness of learning and Sagredo's sharpness of wit, with their kind permission I should wish to leave without hearing any more, as it would appear to me an impossible feat to contradict such palpable experiences. And without hearing any more, I should like to cling to my old opinion; for it seems to me that if, indeed, it is false, it may be excused on the grounds of its being supported by so many arguments of such great probability. If these are fallacies, what true demonstrations were ever more elegant?

3

SALV. Now tell me: If the stone dropped from the top of the mast when the ship was sailing rapidly fell in exactly the same place on the ship to which it fell when the ship was standing still, what use could you make of this falling with regard to determining whether the vessel stood still or moved?

SIMP. Absolutely none; just as by the beating of the pulse, for instance, you cannot know whether a person is asleep or awake, since the pulse beats in the same manner in sleeping as in waking.

SALV. Very good. Now, have you ever made this experiment of the ship?

SIMP. I have never made it, but I certainly believe that the authorities who adduced it had carefully observed it. Besides, the cause of the difference is so exactly known that there is no room for doubt.

SALV. You yourself are sufficient evidence that those authorities may have offered it without having performed it, for you take it as certain without having done it, and commit yourself to the good faith of their dictum. Similarly it not only may be, but must be that they did the same thing too—I mean, put faith in their predecessors, right on back without ever arriving at anyone who had performed it. For anyone who does will find that the experiment shows exactly the opposite of what is written; that is, it will show that the stone always falls in the same place on the ship, whether the ship is standing still or moving with any speed you please. Therefore, the same cause holding good on the earth as on the ship, nothing can be inferred about the earth's motion or rest from the stone falling always perpendicularly to the foot of the tower.

Simp. If you had referred me to any other agency than experiment, I think that our dispute would not soon come to an end; for this appears to me to be a thing so remote from human reason that there is no place in it for credulity or probability.

Salv. For me there is, just the same.

Simp. So you have not made a hundred tests, or even one? And yet you so freely declare it to be certain? I shall retain my incredulity, and my own confidence that the experiment has been made by the most important authors who make use of it, and that it shows what they say it does.

Salv. Without experiment, I am sure that the effect will happen as I tell you, because it must happen that way; and I might add that you yourself also know that it cannot happen otherwise, no matter how you may pretend not to know it—or give that impression. But I am so handy at picking people's brains that I shall make you confess this in spite of yourself.

Sagredo is very quiet; it seemed to me that I saw him move as though he were about to say something.

Sagr. I was about to say something or other, but the interest aroused in me by hearing you threaten Simplicio with this sort of violence in order the reveal the knowledge he is trying to hide has deprived me of any other desire; I beg you to make good your boast.

Salv. If only Simplicio is willing to reply to my interrogation, I cannot fail.

Simp. I shall reply as best I can, certain that I shall be put to little trouble; for of the things I hold to be false, I believe I can know nothing, seeing that knowledge is of the true and not of the false.

Salv. I do not want you to declare or reply anything that you do not know for certain. Now tell me: Suppose you have a plane surface as smooth as a mirror and made of some hard material like steel. This is not parallel to the horizon, but somewhat inclined, and upon it you have placed a ball which is perfectly spherical and of some hard and heavy material like bronze. What do you believe this will do when released? Do you not think, as I do, that it will remain still?

Simp. If that surface is tilted?

Salv. Yes, that is what was assumed.

Simp. I do not believe that it would stay still at all; rather, I am sure that it would spontaneously roll down.

Salv. Pay careful attention to what you are saying, Simplicio, for I am certain that it would stay wherever you placed it.

Simp. Well, Salviati, so long as you make use of assumptions of this sort I shall cease to be surprised that you deduce such false conclusions.

Salv. Then you are quite sure that it would spontaneously move downward?

Simp. What doubt is there about this?

Salv. And you take this for granted not because I have taught it to you—indeed, I have tried to persuade you to the contrary—but all by yourself, by means of your own common sense.

Simp. Oh, now I see your trick; you spoke as you did in order to get me out on a limb, as the common people say, and not because you really believed what you said.

Salv. That was it. Now how long would the ball continue to roll, and how fast? Remember that I said a perfectly round ball and a highly polished surface, in order to remove all external and accidental impediments. Similarly I want you to take away any impediment of the air caused by its resistance to separation, and all other accidental obstacles, if there are any.

Simp. I completely understood you, and to your question I reply that the ball would continue to move indefinitely, as far as the slope of the surface extended, and with a continually accelerated motion. For such is the nature of heavy bodies, which *vires acquirunt eundo*; and the greater the slope, the greater would be the velocity.

Salv. But if one wanted the ball to move upward on this same surface, do you think it would go?

Simp. Not spontaneously, no; but drawn or thrown forcibly, it would.

Salv. And if it were thrust along with some impetus impressed forcibly upon it, what would its motion be, and how great?

Simp. The motion would constantly slow down and be retarded, being contrary to nature, and would be of longer or shorter duration according to the greater or lesser impulse and the lesser or greater slope upward.

Salv. Very well; up to this point you have explained to me the events of motion upon two different planes. On the downward inclined plane, the heavy moving body spontaneously descends and continually accelerates, and to keep it at rest requires the use of force. On the upward slope, force is needed to thrust it along or even to hold it still, and motion which is impressed upon it continually diminishes until it is entirely annihilated. You say also that a difference in the two instances arises from the greater or lesser upward or downward slope of the plane, so that from a greater slope downward there follows a greater speed, while on the contrary upon the upward slope a given movable body thrown with a given force moves farther according as the slope is less.

Now tell me what would happen to the same movable body placed upon a surface with no slope upward or downward.

Simp. Here I must think a moment about my reply. There being no downward slope there can be no natural tendency toward motion; and there being no upward slope, there can be no resistance to being moved, so there would be an indifference between the propensity and the resistance to motion. Therefore it seems to me that it ought naturally to remain stable. But I forgot; it was not so very long ago that Sagredo gave me to understand that that is what would happen.

Salv. I believe it would do so if one set the ball down firmly. But what would happen if it were given an impetus in any direction?

Simp. It must follow that it would move in that direction.

Salv. But with what sort of movement? One continually accelerated, as on the downward place, or increasingly retarded as on the upward one?

Simp. I cannot see any cause for acceleration or deceleration, there being no slope upward or downward.

Salv. Exactly so. But if there is no cause for the ball's retardation, there ought to be still less for its coming to rest; so how far would you have the ball continue to move?

Simp. As far as the extension of the surface continued without rising or falling.

Salv. Then if such a space were unbounded, the motion on it would likewise be boundless? That is, perpetual?

Simp. It seems so to me, if the movable body were of durable material.

Salv. That is of course assumed, since we said that all external and accidental impediments were to be removed, and any fragility on the part of the moving body would in this case be one of the accidental impediments.

Now tell me, what do you consider to be the cause of the ball moving spontaneously on the downward inclined plane, but only by force on the one tilted upward?

Simp. That the tendency of heavy bodies is to move toward the center of the earth, and to move upward from its circumference only with force; now the downward surface is that which gets closer to the center, while the upward one gets farther away.

Salv. Then in order for a surface to be neither downward nor upward, all its parts must be equally distant from the center. Are there any such surfaces in the world?

Simp. Plenty of them; such would be the surface of our terrestrial globe if it were smooth, and not rough and mountainous as it is. But there is that of the water, when it is placid and tranquil.

Salv. Then a ship, when it moves over a calm sea, is one of these movables which courses over a surface that is tilted neither up nor down, and if all external and accidental obstacles were removed, it would thus be disposed to move incessantly and uniformly from an impulse once received?

Simp. It seems that it ought to be.

Salv. Now as to that stone which is on top of the mast; does it not move carried by the ship, both of them going along the circumference of a circle about its center? And consequently is there not in it an ineradicable motion, all external impediments being removed? And is not this motion as fast as that of the ship?

Simp. All this is true, but what next?

Salv. Go on and draw the final consequence by yourself, if by yourself you have known all the premises.

Simp. By the final conclusion you mean that the stone, moving with an indelibly impressed motion, is not going to leave the ship, but will follow it, and finally will fall at the same place where it fell when the ship remained motionless. And I, too, say that this would follow if there were no external impediments to disturb the motion of the stone after it was set free. But there are two such impediments; one is the inability of the movable body to split the air with its own impetus alone, once it has lost the force from the oars which it shared as part of the ship while it was on the mast; the other is the new motion of falling downward, which must impede its other, forward, motion.

SALV. As for the impediment of the air, I do not deny that to you, and if the falling body were of very light material, like a feather or a tuft of wool, the retardation would be quite considerable. But in a heavy stone it is insignificant, and if, as you yourself just said a little while ago, the force of the wildest wind is not enough to move a large stone from its place, just imagine how much the quiet air could accomplish upon meeting a rock which moved no faster than the ship! All the same, as I said, I concede to you the small effect which may depend upon such an impediment, just as I know you will concede to me that if the air were moving at the same speed as the ship and the rock, this impediment would be absolutely nil.

As for the other, the supervening motion downward, in the first place it is obvious that these two motions (I mean the circular around the center and the straight motion toward the center) are not contraries, nor are they destructive of one another, nor incompatible. As to the moving body, it has no resistance whatever to such a motion, for you yourself have already granted the resistance to be against motion which increases the distance from the center, and the tendency to be toward motion which approaches the center. From this it follows necessarily that the moving body has neither a resistance nor a propensity to motion which does not approach toward or depart from the center, and in consequence no cause for diminution in the property impressed upon it. Hence the cause of motion is not a single one which must be weakened by the new action, but there exist two distinct causes. Of these, heaviness attends only to the drawing of the movable body toward the center, and impressed force only to its being led around the center, so no occasion remains for any impediment.

1   Which of the three characters do you think is representing Galileo's views?
2   What was Galileo trying to prove in these passages?
3   How did Galileo relate the case of a ship moving in calm seas to the discussion of falling bodies?

1   The orthodox Aristotelian position is argued by Simplicio (so christened because of this character's esteem for Simplicius, the sixth-century commentator on Aristotle's writings). Sagredo represents the open-minded attitude of an intelligent layman. Salviati explicitly adopts the Copernican position and voices Galileo's own opinions. Just as Galileo had done in Rome, Salviati in the *Dialogue* repeats the arguments against Copernicus drawn from Aristotle and Ptolemy with fluency, develops them and finally disposes of them.

2   Galileo's chief concern here is to establish the impossibility of proving that the earth does not move. He was replying to the arguments against a moving earth which had prevailed since antiquity. According to these, a spinning earth would hurl buildings, animals, etc. into space, birds and clouds would be left behind in the air, projectiles thrown vertically up would never return to earth at the thrower's feet, and similarly stones dropped from a tower would land some distance towards the east of the tower, since the earth was revolving while the stone was in the air.

All of these arguments rested on Aristotle's physics which was based on common sense. Although Galileo had undermined Aristotle's cosmology, the physics on which it was based was still intact. So long as it was believed, following Aristotle, that it was natural for a body to move in *one* particular way: up or down or in a circle, that a body could remain in motion only if there was an associated mover, pushing or pulling, and that all movements were referred absolutely to a central point of the universe, taken to be the earth, no alternative cosmology based on a moving earth was possible. In other words, in order to establish the Copernican system, it was essential to create an entirely new physics. That is why Galileo's *Dialogue* is much more concerned with physics than astronomy. Much of his book deals with the motion of bodies on the earth.

Galileo replied to the commonsense objections that it was impossible to tell from the fall of a stone or from other experiments with projectiles, whether the earth was in motion or at rest. The result would be the same in each case—the stone would land at the foot of the tower. For he argued that if the earth rotated, this motion would belong also to the stone at the top of the tower, as well as to observers at the earth's surface. In the falling stone this circular motion would be combined with a rectilinear motion. Only the latter would be perceptible to the earth's inhabitants who shared in the rotation of the earth. Similarly, birds flying from a tree would also possess the earth's motion and would never be left behind by the earth. Nor would objects on the earth be hurled away.

Suppose, Salviati said in another passage, that an observer was in a cabin below deck. Let there be flies, butterflies, a bowl of water with swimming fish, and an inverted bottle of liquid from which drops fall. Could the observer tell if the ship was moving or at rest? Provided the ship was moving uniformly, Salviati said there would be no clue to allow the observer to decide. On a moving ship, the fish in the bowl would swim to opposite sides of the bowl in the same way as before, the flies and butterflies would not tire in trying to keep up with the ship, the drops of liquid would fall to the same place even though the ship moves while they are in the air, nor would the observer have to make larger jumps towards the stern than to the prow to cover the same space as on a stationary ship. Salviati explained that the reason for this state of affairs was that the observer, the bowl of fish, the butterflies, the inverted bottle and the air in the cabin all shared in the motion of the ship. No more than any human on a moving earth, the observer below deck was unable to judge from any experience that he was not in motion. Uniform motion and rest were equivalent states for the observer. The idea of motion as a *state*, which was important in Galileo's physics, contrasted with Aristotle's conception of motion as a process.

In the extracts referred to, and at greater length in his other works, Galileo constructed a new physics and so cleared the way for the acceptance of a moving earth. In contrast to the idea that objects on earth are naturally at rest and motion is to be explained by the constant presence of a moving force, Galileo asserted that terrestrial bodies could persist in motion indefinitely without the intervention of a force. This type of physics is called *inertial* physics and it lies at the basis of modern physics which dates from Galileo.

As you can see at the end of the final extract, Galileo considered an imaginary experiment involving perfectly spherical balls in motion on a perfectly smooth plane. It would be impossible to achieve these ideal conditions in any actual experiment because of the intervention of friction and imperfections on the spheres. However, this ability to abstract from the conditions of the real world played an essential part in Galileo's formulation of a new science of motion. Contrary to his orthodox Aristotelian beliefs, Simplicio is persuaded to admit that a sphere set in motion on a smooth plane of unlimited extent would continue to move for ever.

3   According to Galileo the closest approach to the realization of these conditions was the movement of a ship over a calm sea. Like the ball moving on an untilted surface, the ship was at a constant distance from the centre of the earth. Neither would have a tendency to accelerate like a falling body (or decelerate in the case of an object thrown up) since gravity was acting constantly. In other words Galileo's conception of the way bodies persist in motion was related to the distance of the body from the earth's centre. Consequently motion could persist only if it was in a circle. The correct law of inertia that undisturbed bodies continue in a state of rest or in a straight-line motion was never formulated by Galileo. His inertial physics still attributed some special role to the earth, even though he had removed it from the centre of the universe.

4.2.1   What purpose have the empty stellar spaces?

On the 'Third Day' an interesting theological discussion occurs in connection with the vastness of the Copernican universe. Simplicio agrees that God might have created the world on a scale beyond our comprehension, but asks why he should have left an immense empty space between Saturn and the stars. For nothing in Nature was in vain; yet these spaces seemed useless. Salviati replied that while humans lacked nothing of God's attention, it was arrogant for us to expect everything in the universe to be of use to us. Sagredo qualified this with the remark that limitations in our knowledge prevented us from understanding how such empty space might be of service to humans as well as knowing for certain if the space really was empty. Had not the telescope revealed unknown bodies in the heavens? We will find that these associated ideas—the purposefulness of all things in the created world, man's arrogance in supposing all was created for his benefit and in expecting to discern the totality of God's intentions—recur in the thought of later centuries.

4.2.2 Galileo's proof of the earth's motion from the tides

So far Galileo had argued that no observation of an event on earth was sufficient to establish that the earth was at rest, because things would look the same whether the earth moved or not. He then proceeded in the discussions of the 'Fourth Day' to present his capital proof for the mobility of the earth, based on the phenomena of the tides. This was one of his favourite arguments for the Copernican system, and with it he brought the *Dialogue* to a climax.

It was well known that the ebb and flow of the tides varied with time in three different ways: a rise and fall at intervals of about six hours each day, the principal effect, and small monthly and annual variations. Galileo set out to establish that tides occur if, and only if, the earth is moving.

The oceans and seas on earth were to be regarded as expanses of water in giant containers. What would be the effect on these vast stretches of water if their container, the earth, was in motion? Would this motion be detected in the behaviour of a fluid, free to move independently? Not if the containing vessel moved uniformly, for then Galileo said the liquid would simply assume the vessel's motion. But if the earth moved irregularly, with accelerations or retardations, the waters would rise and fall, in other words tides would be observed. Salviati illustrated this with reference to one of the barges which carried water across the lagoon to Venice. Suppose, he said to his companions, that this barge runs aground or hits an obstacle and is slowed down or stopped. The water in the barge will continue to have some impetus and will run towards the prow of the vessel, where it will rise, and sink at the stern. The reverse would happen if the barge accelerated. Provided the acceleration or retardation of the vessel was not so great as to move the whole of the contained water violently, the rise and fall of the water would only be perceptible towards the extremities of the barge.

Salviati said that the action of the barge on the water was like that of the earth on the seas. However, he had yet to convince his friends how the earth could be said to move irregularly, since the postulated annual and diurnal motions were uniform. Salviati drew a diagram in which ABCD represents the earth, centre E. The uniform daily rotation of the earth occurs in the direction A→ B→ C→ D. The uniform annual revolution of the earth is represented by the larger circle, centre K. Salviati said the combination of the two motions gave a resultant irregular motion to the earth. He argued that at the part of the earth A, the two motions were in the same direction and so the result was greater than either, while the part of the earth at C was moving slower, since the annual and daily motions were in opposite directions. At C, as in a barge that was slowing down, the waters would rise; at A, on the other hand, the waters would fall.

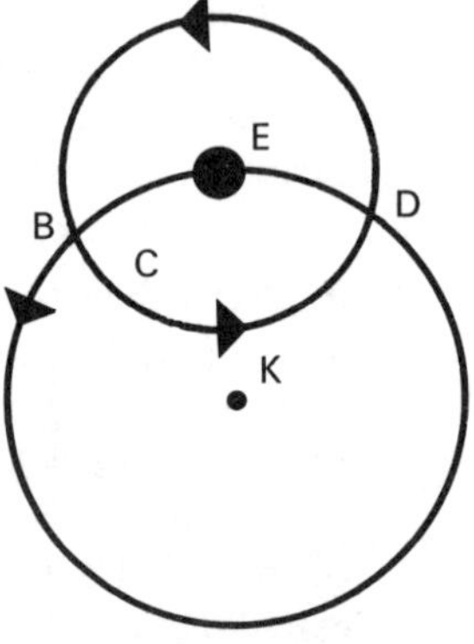

As Galileo appreciated, this theory predicted two tides at intervals of twelve hours, instead of the observed four at six-hourly intervals. He therefore introduced the *ad hoc* hypothesis that further oscillations in the water occurred as a result of variations in the depth of water and other secondary causes. The movement of bodies is described in relation to a definite reference point, for example another object regarded as stationary.

What reference points were employed by Galileo in dealing with the motion of the waters on the earth and with the motions of the earth? Is there anything wrong in this treatment?

Logically there is a fallacy in this argument, which left Simplicio not entirely convinced. It lies in the failure to refer motion to the same reference point. While the orbital motion of the earth is actually considered relative to the fixed stars, the motion of the water is considered relative to the earth. In addition Galileo rejected the theory held by Kepler and others, and later shown to be correct, that the cause of the tides was to be attributed to the direct action of the sun and moon. Galileo regarded this as unscientific thinking, and so restricted the principal cause of the phenomenon to a local, terrestrial nature. It should be pointed out that the rotation of the earth does produce secondary effects on the tides, as do the different ocean depths. So to this extent Galileo's theory had elements of truth.

Galileo's principal argument failed to carry conviction. His contemporaries had continued to regard apparent changes in the positions of the fixed stars as the criterion for deciding on the earth's motion. But like Copernicus before, Galileo could only repeat that these stellar movements were imperceptible. Salviati expected these to be detected in the future by 'extremely accurate observations'. However, it was not until 1838 that an astronomer was able to observe stellar parallax, and not until then was there a proof of the earth's motion, a basic postulate of the Copernican system, in that the motion of the earth relative to the stars was demonstrated.

4.2.3 Simplicio's closing speech

In one of the closing speeches, it is Simplicio who presents Pope Urban's argument and almost in his own words:

> As to the discourses we have held, and especially this last one concerning the reasons for the ebbing and flowing of the ocean, I am really not convinced; but from such feeble ideas of the matter as I have formed, I admit that your thoughts seem to me more ingenious than many others I have heard. I do not consider them true and conclusive; indeed, keeping always before my mind's eye a most solid doctrine that I once heard from a most eminent and learned person, and before which one must fall silent, I know that if asked whether God in His infinite power and wisdom could have conferred upon the watery element its observed reciprocating motion using some other means than moving its containing vessels, both of you would reply that He could have, and that He would have known how to do this in many ways which are unthinkable to our minds. From this I forthwith conclude that, this being so, it would be excessive boldness for anyone to limit and restrict the Divine power and wisdom to some particular fancy of his own.

To which Salviati replies:

> An admirable and angelic doctrine, and well in accord with another one, also Divine, which, while it grants to us the right to argue about the constitution of the universe (perhaps in order that the working of the human mind shall not be curtailed or made lazy) adds that we cannot discover the work of His hands. Let us, then, exercise these activities permitted to us and ordained by God, that we may recognize and thereby so much the more admire His greatness, however much less fit we may find ourselves to penetrate the profound depths of His infinite wisdom.

Does this reply by Salviati fit in with Galileo's views expressed in the *Letter to the Grand Duchess Christina*?

Salviati's words can only be sensibly interpreted by assuming that Galileo is speaking tongue in cheek. It is a reply to the position consistently taken by Rome that hypotheses in astronomy were not to be regarded as true accounts. The reason for this doctrine in the case of the tides was that God's omnipotence would be denied by a necessary demonstration purporting to show that there was only one possible way to produce the phenomenon concerned. Notice also the check imposed on the intellectual pretensions of mortals.

While this doctrine is called 'admirable and angelic' it is given the same status as the view that Man can never discern the structure of the world created by God, a view that Galileo certainly did not share. His earlier impassioned statements on the realistic approach of the astronomer, and his conviction that the human intellect was given by God in order to discern the laws of the created universe, as put forward in the *Letter to the Grand Duchess*, are completely at variance with the literal sense of Salviati's reply. Notice the ironical parenthesis suggesting the goal of human reason to be a mere safeguard against laziness of the mind.

The *Dialogue* ended as it had begun—in irony. In the Preface, Galileo said he would bring together all the arguments he could find in favour of the Copernican system to show the world that Rome had not acted·in ignorance in condemning the doctrine, for all of the arguments had been made known to the Church. He promised to present Copernican theory as a mathematical caprice. Yet no one reading the book could fail to see that the author took the Copernican side of the debate because he believed it to be physically true.

5   The Trial of Galileo

5.1   Proceedings begin

The Pope was furious. He felt that his instructions had been ignored, and worse, that Galileo had made a fool of him by putting his words into the mouth of Simplicio. It was too late to suspend sales of the book—the Florentine Inquisitor gave an order to this effect, but all copies had been sold. The licensing authority, Riccardi, was strongly rebuked by the Pope, though it was felt that he too had been deceived into giving the licence to print. Galileo too was angry. Had he not altered his work to comply with official instructions? Had he not been granted permission to publish? And now he heard that his book was more harmful to the Church than the writings of Luther and Calvin! There seems little doubt that in his enthusiasm for the Copernican position, Galileo misjudged the state of existing opinion.

The Florentine Inquisitor then called at Galileo's house with a summons to go to Rome within thirty days. Although he was offered asylum in Venice, Galileo followed the advice of the Grand Duke of Tuscany and went to Rome, staying at the Tuscan embassy. He was ill and almost seventy years old.

Galileo was to be put on trial before the Inquisition. During the proceedings he was confined in comfortable rooms, far superior to the prisons of the Holy Office which might have been used. The hearings began on 12 April 1633.

5.2   The suspect minute

In the interrogation conducted by the Dominican Commissary of the Inquisition, Galileo was asked to repeat what he had been told by Cardinal Bellarmine in 1616. Galileo replied that he had been ordered not to hold or defend the Copernican opinion, and produced Bellarmine's certificate. The Inquisitor then alleged that this was not correct and read from the Vatican files:

Friday, the twenty-sixth. At the place, the usual residence of the Lord Cardinal Bellarmine, the said Galileo, having been summoned and being present before the said Lord Cardinal, was in presence of the Most Reverend Michelangelo Segizi of Lodi, of the Order of Preachers, Commissary-General of the Holy Office, by the said Cardinal, warned of the error of the aforesaid opinion and admonished to abandon it, and immediately thereafter, before me and before witnesses, the Lord Cardinal being still present, the said Galileo was by the said Commissary commanded and enjoined, in the name of His Holiness the Pope and the Whole Congregation of the Holy Office, to relinquish altogether the said opinion that the Sun is the centre of the world and immovable and that the Earth moves; nor further to hold, teach, or defend it in any way whatsoever, verbally or in writing; otherwise proceedings would be taken against him by the Holy Office; which injunction the said Galileo acquiesced in and promised to obey. Done at Rome, in the place aforesaid, in the presence of R. Badino Nores, of Nicosia in the Kingdom of Cyprus, and Agostino Mongardo, from a place in the Abbey of Rose in the diocese of Monte-pulciano, members of the household of said Cardinal, witnesses.

Compare this statement with that given in Bellarmine's certificate (see Section 3.4). Here you are examining two historical documents which give two different accounts of the same event—Galileo's interview with Cardinal Bellarmine on 26 February 1616. In this situation a historian would carefully examine the wording of the documents, ask himself whom the documents were intended for and how they originated, and decide on their reliability.

What is the essential difference in the two versions on the Cardinal's instructions to Galileo?

In the certificate issued by the Cardinal it is stated that Galileo was instructed not to defend or hold the Copernican opinion. We know the conditions under which this document was issued—it followed Galileo's request for an official statement to dispel the rumours that he had been punished by the Inquisition. The second document states that in addition Galileo was commanded *not to teach* or defend this doctrine *in any way*, and agreed to obey.

Galileo had no recollection of these additional restrictions, and none of his actions since the interview indicates that he was aware of any such impediments. The document concerned is in the form of an administrative minute. It is curious that it is not signed by the notary, before whom the events were alleged to have taken place, or by Galileo.

The authenticity of this minute was challenged by some historians in the late nineteenth century, alleging that it had been planted in the Vatican file in 1632 in order to frame Galileo. The evidence, which was not conclusive, was based on the examination of the manuscript with a magnifying glass. In the 1870s von Gebler reported on his study of the Vatican file dealing with the trial. He found a collection of documents, tattered at the edges and paginated. Some of the papers were just annotations on the decrees relating to the trial. It was clear that this mixed collection was an incomplete historical source, since the papal decrees issued after the end of the trial were not included. At first von Gebler had inclined to the view that the minute describing the injunction was a later forgery. But soon the evidence of the file forced him to reject this, and he decided that the document must be accepted as one written in 1616. The reasons he gave are instructive, since they show something of the way a historian goes about his work. To begin with the handwriting and ink of the suspect document was indistinguishable from that of other documents in the file of undoubted authenticity. In addition all the paper on which the documents of the Holy Office were written in 1616 carried the same watermark—a dove in a circle, and this is no longer found on paper of 1632; yet this watermark was seen by von Gebler on the suspect document. The possibility that the document was written in 1632 on old blank sheets dating back to 1616 was ruled out because of the identity of the handwriting with the earlier documents. Nor could the theory be accepted that a forger cut out the original document, concealing the cut by folding the edge of the paper and then introducing a new sheet with forged wording. This was ruled out by the manner in which the pages were numbered and joined together. The suspect document was the second page of an authentic document. Von Gebler did find that some pages had been cut away leaving strips, but the pages left belonged to complete documents. In any case, would a forger have been clumsy enough to reveal his work by leaving cut strips of paper?

The conclusion that the Vatican file had not been tampered with was confirmed in 1927 when X-rays were used to examine the suspect page. While it is now accepted that the document was indeed written in 1616, it is not known who was responsible for its wording. Various attempts have been made to reconstruct its origin, but they are speculative, and have to be separated from the established facts. The generally accepted view is that the document did not give a true account of the Galileo–Bellarmine interview. Santillana's explanation is that an official present at the interview was disgusted with the easy way Galileo had been let off, and in confidence asked an assistant to insert a stiffer account of the proceedings in the Vatican file. Another leading student of Galileo, Stillman Drake, has suggested that the stronger instructions represent the opinions of Dominicans present at the interview, and that during their expression the Cardinal whispered to Galileo not to pay attention to what was being said.[1] On the other hand Koestler thinks it most probable that the 1616 minute is an accurate statement, and that Bellarmine's certificate represents a watering-down of the text of the injunction, carried out in order to spare Galileo's feelings and put an end to the dispute.

The status of the document is of crucial importance to the assessment of Galileo's trial, since it was the chief means by which Galileo was judged guilty.

5.3 Judgement, sentence and repercussions

A few days after his first interrogation, the three counsellors appointed to examine the text of the *Dialogue* issued their report. They had no difficulty in showing that Galileo had not simply discussed the Copernican theory as a mathematical hypothesis, but had held, defended and taught it as a physically true account. In an attempt to get a settlement out of court, the Commissary General of the Inquisition paid Galileo a visit, advising him that he would be treated leniently if he admitted his errors. This Galileo soon did in a statement that pride had led him to conjure up convincing novel arguments for the Copernican doctrine. He added that to show his disbelief in this theory he was prepared to add one or two other 'days' to the *Dialogue* to disprove it more effectively. This must have been a bitter retraction for Galileo. He had no sensible alternative but to throw himself on the mercy of the court. It is significant that with this insincere avowal Galileo continued to maintain that the 1616 information was news to him.

Worse was yet to come. The affair could have ended there, but instead Galileo was to be humiliated further. This move, which seems to stem from the Pope's enmity, required Galileo to be interrogated further under a nominal threat of torture (this was not a real threat, as it was illegal to use torture on a man of Galileo's age). Galileo was to declare which astronomical system he adhered to. Again he retracted, stating untruthfully that he accepted Ptolemaic astronomy, and abandoned the Copernican system. Two days later Galileo knelt in a white shirt of penance before his judges in the Dominican Convent of Santa Maria Sopra Minerva. Sentence was passed, prohibiting the *Dialogue*, condemning Galileo to life imprisonment and requiring him to repeat penitential Psalms once a week for three years. Galileo signed his abjuration. Although the Pope did not sign the sentence or ratify it he was clearly behind the action taken. However at no time in the Galileo case did Urban declare that he was speaking *ex cathedra*. So the question of papal infallibility does not arise here. (The infallibility of popes was not asserted as a dogma until the nineteenth century.)

Galileo had been sentenced with the help of a dubious document on a charge of heresy; yet the Copernican doctrine had never been declared such by a Pope. Justice had not been done. Later Galileo looked back bitterly on the trial as a triumph for ignorance, fraud and deceit.

Imprisoned at his farm in the hills surrounding Florence, Galileo's sight was failing. Nevertheless it was here that he wrote his most important scientific work, his *Discourses Concerning Two New Sciences*, which was published in Holland in 1638. Dealing with falling bodies and the path of projectiles, this work laid the foundations for modern kinematics. He died in 1642.

Blame for the Galileo tragedy has been variously attributed to the prevarication and vindictiveness of Pope Urban, the conservatism of Cardinal Bellarmine, the passivity of some Jesuits and the intrigues of others, the malice of Caccini, the conspiracies of the 'pigeon league', and to the obstinacy and cantankerous nature of Galileo. A question of greater significance is why the conflict should have occurred at all? The Galilean affair was an unfortunate and dramatic manifestation of the crisis in Western thought created by the growth of evidence against a view of the

[1] This view is mentioned in a footnote in James Brodrick's *Galileo, The Man, his Work, his Misfortunes*, G. Chapman, 1964, p. 134.

universe which had been established for centuries. With the development of modern science, a confrontation with the established Aristotelian beliefs about the world was bound to occur. Galileo was attempting to destroy the existing world picture. The fundamental issue was raised of the natural philosopher's right to pursue his research independently of received opinion, including that of the Church.

The repercussions were such that for over a century Catholic natural philosophers felt obliged to present their astronomical and cosmological treatises in the guise of a mathematical hypothesis. Not until 1757 was the anti-Copernican decree revoked by Pope Benedict XIV. However, Galileo's *Dialogue* remained on the Index until 1831. Some official redress was also made by Pope Leo XIII in his 1893 encyclical which endorsed the approach to the Scriptures taken by Galileo.

Further Reading

Discoveries and Opinions of Galileo, translated with introduction and notes by Stillman Drake, Doubleday Anchor paperback, 1957.

Galileo Galilei, *Dialogue Concerning the Two Chief World Systems—Ptolemaic and Copernican*, translated by Stillman Drake, University of California Press paperback, 1962.

Giorgio de Santillana, *The Crime of Galileo*, Chicago, 1955; Heinemann Education, Mercury paperback edition, London, 1961.

J. J. Langford, *Galileo, Science and the Church*, University of Michigan Press, revised edition 1971.

Alexander Koyré, *Metaphysics and Measurement: Essays in the Scientific Revolution*, Chapman and Hall, 1968, pp. 1–43.

Acknowledgements

Grateful acknowledgement is made to the following sources for material used in this unit:

Text

Doubleday & Company Inc. and the author for excerpts from *Discoveries and Opinions of Galileo* by Stillman Drake. Copyright © 1957 by Stillman Drake. Reprinted by permission of Doubleday & Company Inc; University of California Press for Galileo Galilei, *Dialogue Concerning the Two Chief World Systems—Ptolemaic & Copernican* (trans. Stillman Drake). Originally published by the University of California Press; reprinted by permission of The Regents of the University of California.

Illustrations

Figures 1 and 3: Science Museum, Crown Copyright; *Figure 2:* Ronan Picture Library; *Figures 4 and 5:* Harvard University Press; *Figure 6:* Mansell Collection; *Figures 7 and 8:* British Museum.

Notes

Notes

Notes

Science and Belief: from Copernicus to Darwin